ROUTLEDGE LIBRARY EDITIONS:
20TH CENTURY SCIENCE

T0227536

Volume 1

SCIENCE IN THE CHANGING WORLD

BOUND WITH

SCIENCE AT YOUR SERVICE

SCIENCE IN THE CHANGING WORLD

THOMAS HOLLAND, H. LEVY, JULIAN HUXLEY,
JOHN R. BAKER, BERTRAND RUSSELL,
ALDOUS HUXLEY, HUGH I'A. FAUSSET,
HILAIRE BELLOC, J. B. S. HALDANE AND
OLIVER LODGE.

Edited by
MARY ADAMS

BOUND WITH

SCIENCE AT YOUR SERVICE

JULIAN S. HUXLEY, EDWARD APPLETON,
GEORGE BURT, LAWRENCE BRAGG,
J.B. SPEAKMAN, JOHN READ, A. O. RANKINE,
NELSON JOHNSON, MICHAEL GRAHAM,
ALBERT PARKER, J. L. KENT, G. L. GROVES
AND E. C. BULLARD

With a Preface by
E. C. BULLARD

Routledge
Taylor & Francis Group

LONDON AND NEW YORK

Science in the Changing World first published in 1933
Science at Your Service first published in 1945

This edition first published in 2014
by Routledge
2 Park Square, Milton Park, Abingdon, Oxfordshire OX14 4RN

and by Routledge
711 Third Avenue, New York, NY 10017

First issued in paperback 2016

Routledge is an imprint of the Taylor & Francis Group, an informa business

British Library Cataloguing in Publication Data
A catalogue record for this book is available from the British Library

ISBN: 978-0-415-73519-3 (Set)
ISBN 13: 978-1-138-98142-3 (pbk)
ISBN 13: 978-1-138-01330-8 (hbk)

Publisher's Note
The publisher has gone to great lengths to ensure the quality of this book but points out that some imperfections from the original may be apparent.

Disclaimer
The publisher has made every effort to trace copyright holders and would welcome correspondence from those they have been unable to trace.

SCIENCE

IN

THE CHANGING WORLD

by

THOMAS HOLLAND
H. LEVY
JULIAN HUXLEY
JOHN R. BAKER
BERTRAND RUSSELL
ALDOUS HUXLEY
HUGH I'A. FAUSSET
HILAIRE BELLOC
J. B. S. HALDANE
OLIVER LODGE

Edited by

MARY ADAMS

LONDON
GEORGE ALLEN & UNWIN LTD
MUSEUM STREET

FIRST PUBLISHED IN 1933

PRINTED IN GREAT BRITAIN BY
UNWIN BROTHERS LTD., WOKING

EDITOR'S INTRODUCTION

THIS book is based on a series of broadcast talks on science, which formed part of a comprehensive symposium on *The Changing World*. In that symposium an attempt was made to reflect the crisis through which the world is passing and to make an analysis of those forces of transformation in science, art, economics, and social life which have been in operation since the beginning of the century. All the speakers were preoccupied with the same general theme and, within each particular field of inquiry, set themselves to answer the same questions, thus achieving for the first time in the history of broadcasting a unity of theme and a continuity of treatment over a considerable period of time.

The dominance of science over the day-to-day lives of our contemporaries gives a special interest and significance to the analysis of the changes which have been brought about by progress in scientific thought. The practical applications of science order our civilization. Science generally enters the lives of ordinary individuals as a mechanical device or a social convenience—a motor-car, a wireless set, or a telephone. The impression is widespread that science is the history of sudden and startling inventions rather than a method of pursuing truth. It is the aim of Professor Levy's contribution in Part I to disclose the fundamental nature of science: that it is a process of systematic trial and error, of frustration and discovery, a laborious construction of instruments, theories, and methods of investigation. Professor Levy defines the scope of scientific inquiry, and stresses the importance of the scientific outlook for the investigation of the motives of human behaviour.

Man's investigation of himself is a significant development of twentieth-century science, and the biologist has a definite contribution to make to any discussion of human nature. Researches into our ancestry, our growth, and our conduct have practical applications to everyday affairs, casting light on urgent social problems, compelling tolerance, and occasionally indicating profitable adjustments. The science of human heredity is beginning to affect the social conscience and to provoke speculations about the biological future of the race. Some of the facts necessary for an appreciation of biological questions are provided in Part II by Dr. John Baker, while Professor

Julian Huxley discusses their meaning in relation to the environment in which man exists.

Finally, in Part III, the civilization which constitutes our environment comes under scrutiny. Men with views as widely divergent as those of Mr. Hilaire Belloc and Professor J. B. S. Haldane examine critically the philosophical, aesthetic, and social implications of scientific progress. It cannot be denied that science has brought material freedom, wealth, and leisure, and liberation from famine and disease. But has science produced these things at the cost of spiritual atrophy and personal servility? Does a machine-made civilization stifle artistic expression, or does it merely transfer its sphere of activity? Under the machine, are all the elements of man's personality able to find harmonious expression? Is it possible to view the ever-moving frontiers of science with equanimity? Is the advance of science inevitable, or does scientific progress contain within itself the seeds of decay? There is a widespread sense of disharmony between the old and the new from which spring endless perplexities and conflicts. Will the fabric of society stand the strain of such swift change? Man's weaknesses have been exposed, and doubts are freely expressed about his inherent capacity to control scientific "progress."

Man is out of place in nature, and some of those who are contributing to the symposium feel that unless some kind of re-orientation occurs he cannot survive. On the other hand, other contributors believe that the remedy for our sickness is not less science but more, that a more scientific understanding of human nature will restore coherency to life, and that the ancient forces of religion, aesthetics, and humanism, will find their place in the modern age. But in that event we must accept the inevitability of science and apply ourselves to the task of understanding the civilization in which it works: there is hope for the future only if we strive to condition it by philosophic forethought and scientific planning.

Many listeners said they wished to have the talks in book form, and their publication may be welcomed by those who took part in wireless listening groups, reminding them of the companionship and discussion which the talks themselves occasioned.

It has been thought desirable to preserve the simplicity of style and the atmosphere of informality which characterize broadcasting, so some of the talks are published without modification. Alterations in others have become necessary, and the editor is responsible for certain rearrangements in their presentation.

The talks were arranged under the auspices of the Central Council for Broadcast Adult Education, and this opportunity is taken of acknowledging the courtesy of the British Broadcasting Corporation in giving permission for their publication.

MARY ADAMS

LINCOLN'S INN

December 1932

CONTENTS

B

PART III

WHAT IS CIVILIZATION?

CONTRIBUTORS

JOHN R. BAKER

University Demonstrator in Zoology, Oxford. Author of *Sex in Man and Animals*, etc.

HILAIRE BELLOC

Author of *Joan of Arc, The Path to Rome, History of England, Essays of a Catholic Layman in England*, etc.

HUGH I'ANSON FAUSSET

Critic, and author of *The Proving of Psyche, The Modern Dilemma*, etc.

J. B. S. HALDANE, F.R.S.

Professor of Biology in the University of London. Late Reader in Biochemistry, Cambridge University; Head of Genetical Department, John Innes Horticultural Institution; Fullerian Professor of Physiology, Royal Institution. Author of numerous scientific papers on human chemical physiology, genetics, natural selection, and other subjects.

SIR THOMAS HOLLAND, F.R.S.

Principal and Vice-Chancellor of University of Edinburgh. President of the British Association, 1929; Rector, Imperial College of Science and Technology, 1922–29. Author of publications on Petrology, Geology, and Anthropology.

ALDOUS HUXLEY

Author of *Antic Hay, Point Counter Point, Proper Studies, Brave New World*, etc.

JULIAN HUXLEY

Honorary Lecturer, King's College, London. Author of *Essays of a Biologist, Essays in Popular Science, The Science of Life* (with H. G. and G. P. Wells), as well as numerous scientific papers.

H. LEVY

Professor of Mathematics, Imperial College of Science. Author of *Aeronautics in Theory and Experiment* and other technical and scientific works as well as *The Universe of Science.*

SIR OLIVER LODGE, F.R.S.

Rumford Medallist; Romanes Lecturer, 1903; Past President of the British Association. Author of *Science and Human Progress* and many other books on philosophy and science.

BERTRAND RUSSELL, F.R.S.

Author of *Introduction to Mathematical Philosophy, The Analysis of Mind, Sceptical Essays, Marriage and Morals, The Scientific Outlook, Education and the Social Order,* etc.

THOMAS HOLLAND

INTRODUCTION

THERE is nothing mysterious or strange about the methods adopted by the worker in Science. In his attempt to find out the nature, and therefrom the history, of the universe, he employs the methods adopted by the ordinary historian. His methods are indeed simpler, for, whilst he is often embarrassed by gaps in the records, he is sure that they have not been tampered with artificially; he knows that every observation has some significance, each is worth recording faithfully as so much positive knowledge.

It seems easy to say that the work of the scientific student consists of the systematic record of facts, their grouping into classes of like kind, the drawing of deductions from established propositions, and their verification by further experiments and observations. But throughout all these apparently simple stages the scientific worker suffers like other people from a natural temptation to form theories, to extend his conclusions back beyond the region of actual observation, and to forecast the future.

During the evolution of the animal world some species found it convenient to utilize trees for their habitat, and so to use the instrument of sight over wider ranges, giving increased opportunities and greater safety. Such devices were often attempted by earlier species without permanently advantageous results. But in a late period of the Earth's history, certain shrew-like animals, with well-developed brains, were in a condition to turn this change of habit to better account, and their descendants benefited by the development of a special brain mechanism for the purpose of correlating the impressions received by sight, as well as through the other various sense channels.

In time descendants arose in which this so-called neopallium grew, until in size it exceeded the remaining total of the central nervous system. It is this neopallium especially that distinguishes us from all other animals: it is this which has given us advantages over the living world in material things: it is that which has brought us all the troubles of an imperfectly controlled imagination.

Before it was brought under discipline by scientific methods, the activities of man's neopallium peopled the sun, the other heavenly bodies and many special manifestations of Nature with spirits, mainly evil in disposition —spirits that had to be appeased by sacrifices; the first-born of the family, the first-born of live-stock, the first-fruits of agriculture, and in a later, more anaemic age, by alms-giving. Among some animal communities practices that are instinctively regarded as anti-social are punished by death; it was left to man, under the domination of an undisciplined neopallium, so to treat unorthodoxy in belief as a capital offence. It is only 331 years since Giordano Bruno was burnt at the stake for preaching that this Earth is not the centre of the universe. This form of punishment is now out-of-date, but milder penalties are still imposed to show that the preconceptions of the neopallium dominate the clear evidence of those senses that guide more simple-minded animals. It is only six years since a Tennessee schoolmaster was prosecuted by the State authorities for teaching evolution as the history of the biological world. And yet, what was at one time a plausible hypothesis, and later adopted as a working theory, is now by a consistent mass of evidence as much a fact in the history of the World as the cross-channel trip of William the Conqueror.

Reactions of this sort may still be numerous among individuals, but are fortunately rare now among civilized governments; so rare indeed that many people find it difficult to believe that the Tennessee incident is fairly indicative of a widespread mental disposition. But it is illustrative of the fight that constantly goes on between what we imagine and what we actually see or otherwise know from tested facts. The primary aim of Science is to obtain such reliable facts, but no one can restrain the constant temptation to offer an interpretation, to form a theory. Indeed, it is from theories so formed that clues are obtained for further experiments and wider observations. Charles Darwin in biological research and Michael Faraday in the physical sciences were outstanding examples of workers whose remarkable success seemed to arise from a dominant tendency to subordinate theory to observation and experiment; and yet both were constantly guided by working theories, both of them realizing always that whilst the theory is the product of one's own imagination, the facts of Nature are of divine origin.

It is useful to recall Faraday's own words:

> The world little knows how many thoughts and theories which have passed through the mind of a scientific investigator have been crushed in silence and secrecy by his own severe criticism and adverse examinations; that in the most successful instances, not a tenth of the suggestions, the hopes, the wishes, the preliminary conclusions have been realized.

And so the history of Science is marked by occasional modifications of theories, not, however, as often, or by changes as revolutionary, as the public are often led to suppose. A short time ago a leading journal published this partial truth: "Science to-day does not hold quite the authoritative position it did, mainly because its fallibility has been exposed by itself. What science says to-day it unsays to-morrow."

The reference is to some of the extensions of

theory due to recent discoveries in physics; but the facts which it is the primary business of Science to unearth remain the permanent heritage of the human race. So far as I know, experimental and observational Science has never had to take a backward step. Even in theory modifications are generally not reversals, but improvements in the direction of refinement and precision. Take the instance of the new physics which is so frequently quoted as revolutionary. During the latter half of the nineteenth century we were happy with the ideas that radiant energy, like light and heat, is conveyed as waves of an unknown medium called the ether; that whilst matter might be changed in physical form it could not be destroyed; that it is composed of elements with distinct and fixed properties.

All scientific men suspected something beyond these ideas, and they remained true for all conditions under which they were tested before 1895. On them we based the manifold material applications of Science, and as such they have never failed. They are still reliable guides in the application of Science. Nevertheless they do not express the whole truth. Our nineteenth-century atoms are still atoms in spite of their inappropriate name; but we find that with new methods they can be dissected as we previously dissected compounds; and we now know too that some of the elements break up into others. Just as Faraday showed that one form of energy can be transformed to another, we now find that, under special conditions, mass can be transformed to energy and energy can reproduce the properties of mass. We have yet to find some unifying law to explain the newly acquired groups of facts. The old laws are still safe guides within all previously known conditions.

Along the journey in the search for more truth we occasionally take a wrong turning; so the animals and plants did in their progress towards more complex types, but, as with them, the end result is real progress. In questions physical we are obviously just entering previously unknown territory; our expectations and our curiosity are stirred to-day more perhaps than they have ever been in the World's history.

In the biological world, too, one sees the approach of solutions so clearly that one feels envious of the younger generation that will take part in the exploratory work. Here we have an Earth that was once certainly unfit for the maintenance of what we call life, an Earth that some day will again be unfit for habitation by living beings. What started life and what constantly pressing hidden influence caused innumerable generations by a process of trial and error to reach the product of a being with a mind capable of exploring other worlds?

When one realizes how limited are the conditions that maintain life, one wonders whether, after all, the "accident" is not unique and that man after all has some mathematical justification for his traditional conceit and self-complacency. Life is possible only in aqueous systems and is maintained by perishable combinations of a few elements; the limiting conditions are dangerously narrow; yet the possible permutations and combinations run to figures comparable to those with which astronomers dumbfound their hearers when they talk of star-distances in light-years. These unstable compounds have been continually reforming themselves out of simpler combinations that exhibit none of their special properties; they have been doing this without interruption from generation to generation for more than 900 million years, building up new forms that have branched off from the main stream ultimately, one after the other, to become extinct; but still the main stream persisted until, following what seems like miraculous escapes from an incalculable number of "accidents," the human mind is produced as a final product. Could all these "accidents" in any other world have possibly resulted in exactly the same kind of product? The Earth as a stellar body is less conspicuous than a sand-grain on the sea-shore, but Man, its highest product, may yet be unique in the Universe.

But I must forsake these speculations, even as questions, or I shall be in danger of straying beyond what Professor Levy will define and illustrate as the legitimate garden of the student of science.

PART I

WHAT IS SCIENCE?

H. LEVY

H. LEVY

1. THE PARADOX OF SCIENCE

THE last century has seen such a development of scientific knowledge that the time is past when any one person can hope to have a detailed understanding of the whole field of science. Around the fireside we may be wiseacres, understanding everything from stainless steel to smoky chimneys, but there is no scientific man who would dare to claim expert knowledge about many different problems. Take that little packed cylinder of tobacco you may be smoking at the moment—I do not know a single scientist who would claim to know all about tobacco-growing, about the manufacture and composition of cigarette paper, and whether smoking is harmful or not. Yet these are only three of the many highly technical questions one might discuss.

Again, in spite of much popular misunderstanding, science is not a definite clearly defined body of knowledge. It is never possible to say that this or that is the last word on any scientific subject. On the contrary, science is continually expanding; it is in a continual state of change. Yesterday matter was thought to be the fundamental stuff of which our universe was made, this morning it was atoms, this afternoon it was electrons, this evening it is something much less definite—a wave—radiation. What will it be to-morrow?

Here is a piece of paper: it has shape and size, a certain chemical make-up, a definite weight. I might be able to describe all these aspects of the paper with great apparent accuracy, and yet I cannot tell you why it is that when I try to lift the paper by one corner the rest of the paper is lifted with it. How is it that all these exceedingly small particles—molecules, atoms, electrons—which science offers as the basic stuff of matter, hold together in the shape of this paper, so that it moves as a whole when I pull one corner? It is amusing to realize that no scientist who values his reputation would assert that he really knows the answer. An adequate reply to this apparently straightforward question might make all our previous descriptions of its weight, shape, and chemical composition look quite different. You see, scientists are trying all the time to upset their own equilibrium. They are continually digging away at their foundations. What anchor-hold on such shifting sands, you may well ask, can science give?

The fact is, of course, that science prides itself on this capacity for change. It is prepared to take every scrap of verified evidence into consideration, whether or not it accords with the personal likes or dislikes of the investigators themselves. It is the solid basis of assured knowledge continually and relentlessly accumulating by this process which provides the anchor-hold of science. That this anchor must be constantly tested is clear, and so the evidence upon which scientific fact rests must be continually examined and overhauled. It is impossible, therefore, to state what science is, at any one time, without describing also the process by which science acquires its facts. We have to realize, moreover, that these facts are collected and interpreted by man. Now man possesses certain limitations which we must not underrate, for they affect his interpretations very profoundly. Man's picture of the world is not like a photograph. True, he handles, at close quarters, the impersonal objects of his world—the "earth, air, fire and water," animals and plants, atoms and electrons—but his observation and understanding of these things depend also on his senses; his interpretation is influenced by his inheritance and by his environment.

When at birth he is plunged into this changing world, he enters into two main heritages. In the first place he has acquired a bodily structure, apparently complete, with bony frame, muscles, and sense organs, a bodily organization which evolved through countless generations from early forms of life. The range and power of his sense organs set limits to what he can see and hear and smell and feel. Scientific instruments—telescopes, microscopes, weighing machines, telephones, and so on—have only comparatively recently extended his powers of perception. He cannot see the bones of his body; he can only feel the position of some of them. An X-ray apparatus, however, can help his eyes and his fingers to do these things. In the same way he is unaware, unaided, of the beautiful colours in the inside of his body. The soft tread of a fly, and the murmur of dust being deposited all around him, is not distinguishable by his ear, but a microphone could enlarge these sounds to the patter of hail. Such a magnification makes one realize how limited are our direct powers of perception.

In the second place man inherits a social environment. Most of us are born into a home, have school companions, friends, and acquaintances; we are members of a church or a trade union or a parish. We find ready-made institutions, books which have been read and laws which have been obeyed for hundreds of years—a mass of established tradition and belief. All these influences surround us from birth with rules of conduct, social taboos and prohibitions, shaping the greater part of our behaviour and colouring our thoughts until we die. Our attitude towards our parents or our children, to individuals in other social classes, towards religion, politics, and so on, is more or less determined by this social environment. We have customs and beliefs which are scarcely more than historical relics of our savage origin. We have taboos about food, about thunder. We are still very close to primitive man in outlook, temperament, and social background—a fact which becomes evident in times of danger and of great excitement. It is easier to see the caveman in others than in ourselves. Only about ten thousand generations, after all, separate us from our savage ancestors. As I sit here in Central London I can imagine my parents and grandparents and all my ancestors standing in succession one behind the other stretching southwards to the sea. Only my grandfather, my father, and myself are aware of the existence of this procession, and we do not know its full significance. But looking back along the line we can see that long before the procession has reached the outskirts of Greater London our ancestors have become naked wandering savages. Throughout Surrey and Sussex they become increasingly ape-like, and by the time the sea is reached they can hardly be distinguished from the tailless apes. Civilization is almost within earshot. Our ancestors were agriculturists and metal workers on the threshold of this building, and in the passage outside they are reading books and presently suggesting that the way to find out about the world is to make experiments with it.

So here you see man's historical background—a background which we must never forget when we are discussing the meaning of our science, our religions, and our philosophies. At each successive stage in history man is apt to regard his explanation of the world as final and complete, ignorant of the fact that his explanation is little more than a reflection of the ideas and beliefs of his own particular scene in this historic pageant.

When I set out to tell you what science is, or indeed when anyone sets out to tell you *really and truly* what anything is, as if the explanation were the last word on the subject, you will, I hope, find yourself doubting its finality. All explanations must be examined in their evolutionary setting. I stress this point because our newly found ability to look at ourselves and our ideas in this way is, to my mind, one of the most significant changes that have been brought about by the science of the last century. That change is one of the greatest contributions of science to education. Whether educationists have, in their field, exploited this fact to the fullest possible extent is another matter.

There is another reason why my task of defining science is difficult. The traditional picture of the scientist as a bespectacled individual, so immersed in his researches in his laboratory that he is unconscious of the havoc his work is producing in the outside world, is not entirely a caricature. It is, in fact, often true. The scientist is usually so absorbed in his tiny specialized field that he rarely has time or opportunity to think about the social effect of his labours or to look in perspective at the movement of which he is a unit. Science is a very absorbing pursuit, and it may be that the mental concentration which it requires provides an escape from the trials of everyday life. Nevertheless the time has gone when the scientist could legitimately separate himself from the rest of his fellow-men in the belief that his scientific interests were his own and that they affected no one. It is true that Faraday's early studies of electricity were primarily of laboratory interest. Later on industrialists saw in the practical application of his work a possible source of fresh profits. But this generation, living in a world of electrical devices and of industrial disorganization, is being taught by bitter experience that it is disastrous to keep science and its industrial applications in water-tight compartments. The scientist and his work cannot be separated from the rest of his changing universe. Science has social roots and social consequences.

We are all of us continually making this false separation even with the most everyday things. For example, you have a pencil or a cigarette in your hand, and you easily think of it as a thing by itself, a separately existing object. But is it? It is a cigarette—in your hand. Your hand is attached to your body, your body sits in a chair, the chair is on the floor, the floor is in a building, the building in on the earth, the earth is part of an assembly of planets careering round the sun, and the sun and our solar system are only part of other systems.

The point is, that we easily separate off for examination tiny separate fractions—the solar system, the earth, the building, the floor, the body, the hand, the cigarette. But notice that we only do this for simplicity's sake, to make our examination more easy. The separation is nevertheless quite artificial. We cut our cigarette out of the universe as if it were a separately existing entity. But there is no such thing as a cigarette *in itself*. No one has ever seen one, it is a figment of the imagination, a pure abstraction. In the same way it is not easy to make ourselves remember that this cigarette we have so boldly plucked from the rest of the universe is now different from what it was a moment ago, and from what it will be in a moment to come. There is a continuous process of change going on, so that to say our cigarette preserves its identity through time and space is also a pure abstraction. Of course it is easy to appreciate this point when a burning cigarette is changing rapidly before our eyes, but the same considerations apply to everything else we so easily call an object—a piece of iron, a stone, the planet on which we live. The world you and I perceive is a world of perpetual change of which we are an integral part.

I stress this point because I want to show you that our common-sense way of looking at the world, regarding it as composed of a number of separate objects, may not tell us the whole story. Many of the growing-pains of science have arisen from this fact. Once an object has been separated off and given a name we seem to expect the object to persist unchanged because the name persists.

What does it matter in practice? one may ask. Very little for most purposes. We do, in fact, live the greater part of our lives as if the objects we handle were permanent and separate things. Fortunate it is for science that we do so, for much of its framework is in practice built round this conception of permanence. In the newer physics, however, these ideas matter a great deal. Only a few years ago the indestructibility of matter and the indestructibility of energy were accepted almost as religious beliefs in science, so accurately did matter and energy maintain their separateness in practice. Then came the newer knowledge given us by the study of electricity, and these hard and fast ideas of permanence had to be abandoned when radium and similar substances were found to discharge tiny electrified particles. In the face of an extended experience the old abstractions of permanent separate matter and permanent separate energy broke down. Matter in certain circumstances dissolved into energy.

These are not mere manufactured difficulties. The perplexities stand out as soon as one attempts to lay down a basis for accurate knowledge, and if one ignores them one builds on a basis of falsehood. The answer to the question "What is Science?" is then no mere definition or form of words. It will be found by studying the scientist at work. Outside the realm of pure mathematics there is little that can be described by mere definition. We can tell what anything is only by examining the process which exposes it, and by studying it in relation to the wider processes of which it is a part. We do not define things into existence.

I propose then to set out the position we have reached so far:

(1) We must regard any knowledge we acquire about the world in the setting of man's historical evolution.

(2) Both we ourselves and the world of which we are a part are in a continual state of change. It is untrue to say that there is nothing new under the sun. In a sense there is something different under the sun every moment.

(3) These world changes penetrate to us through our sense organs. Not only are these limited in power and range but they also have an evolutionary history. Tools and scientific instruments are inventions for extending their powers.

(4) Science studies the changing world by the method of abstraction. The scientist separates off from the rest of the universe any object he wishes to study. The method of abstraction sounds difficult, but it is, in fact, the method of ordinary discussion.

May I explain what I mean by the method of abstraction? If you are considering how high a ball will rise when you throw it into the air you are not concerned with the colour or chemical nature of the ball or when and where the ball was made. You are concerned only with its shape, size, and weight, and with wind resistance. Any other sphere with the same shape, size, and weight would do equally well. We have ignored what, for immediate purposes, is irrelevant and have abstracted those things which are significant for the purpose in hand. From this process there emerges a common principle at work, on which one relies in explanation.

An abstraction therefore, although it sounds complicated, is really a simplification. By means of abstraction irrelevancies are stripped away and fundamental likenesses between different objects are thereby disclosed.

An explanation consists in describing complex events in terms of simplified abstractions.

"Well," I can hear some of you grimly remarking, "now we know." "Why is it," you ask, "that scientists use so many unfamiliar words that their talk sounds like a foreign language? Why cannot they bring themselves down to the level of the ordinary man?"

It is not pure perversity: there is a defence. A hen crosses the road and I ask, "Why does that hen cross the road?" A satisfactory answer is, apparently, "Because she wants to get to the other side." I inquire how you know what the hen wants. You cannot tell me that your evidence of what the hen wants to do is derived from the fact that she actually does cross the road, for that would be begging the question. How do you know what the hen wants? You may dislike this persistence, for I suspect you have been transferring to the hen your own private feelings about roads and walking. When I begin to explain, however, how it is that the hen crosses the road, and I use in my explanation words like "external stimuli," "conditioned reflexes," "motor reaction," "visual reception," you begin to abuse me because I am talking a foreign language. The fact is that the whole question of language is a very vital part of the process of explanation and the means of arriving at scientific truth.

The scientist has to distinguish between two kinds of statement. In the first place there are statements about the so-called *external* world (it is not external for we are pieces of it). For example, this room is 20 feet square. This paper is white. That note is E flat. These are statements that can be verified. I can communicate them to you with assurance because I know that you can verify them if you take enough trouble. The verification may be extremely difficult and most of us have to rely upon individuals with special knowledge and elaborate apparatus to verify them for us. For example, you will not be able easily to verify the statement that the speed of light is 186,000 miles per second. You have to rely for the truth of that statement upon the fact that this figure has been arrived at independently by scientists working in many different laboratories. It is with these *public* affairs that science operates, and only those things which can be verified *publicly* are included in the term scientific knowledge.

We may, nevertheless, receive pleasure from a public discussion on a private matter—on literature or on art. These are statements about private feelings. "The wind bloweth where it listeth" is poetry not meteorology. A phrase like "Nature abhors a vacuum" is not a scientific explanation of the reason why a tube with one end closed, from which air has been expelled, immediately fills with water if the open end be immersed in that liquid; and yet the phrase "Nature abhors a vacuum" may be found in text-books even to-day. Science knows nothing about this ill-defined dislike of a vacuum on the part of Nature. Dislike is an expression of personal feeling. Of course, popular language is honeycombed with these expressions, and the language of a scientist appears dull and complicated to a layman because the scientist has to exclude these poetic, colourful, but, for scientific purposes, meaningless descriptions. His explanations must be publicly verifiable.

To devote so much space to these considerations may appear excessive. It is this scientific method of public investigation, however, which has been mainly responsible for the vast changes in civilized life which have become apparent during the last hundred years. It has made ours a different world from that of our great-grandparents. Yet it can hardly be said to be in common practice outside scientific laboratories. Man, in fact, has not yet caught up with his own method of investigation. He is one of the changing objects in this changing world, and he changes slowly. His schools and his laws, and those social institutions which settle so much of his belief and behaviour, still drag slowly behind. He still has vain imaginings, fears, and personal egoisms which colour his discussion and argument.

One need not be learned in scientific matters in order to acquire the point of view I have here outlined. You will find, I suggest, that if you avoid private explanations in discussion, restricting yourself and your friends to public matters that can be verified, truth will acquire a new and cleaner complexion. Motives will be verified not by personal assurances, but by an examination of actual behaviour. Discussions that might have finished in personal bickerings and estrangements may resolve themselves into collective attempts to obtain and examine evidence.

2. SCIENCE IN REVOLT

I HAVE tried to make it clear that the scientist approaches the matter-of-fact world in a manner not really very different from that of the man in the street. The scientist in his capacity as scientist, however, is prepared to talk only about certain things, and to offer as satisfactory only certain kinds of explanation. I have tried to show, too, just how the scientist, and for that matter the layman also, sets about the difficult task of examining the world into which he was born. I have also stressed the fact that the explanatory language of science does not include any reference to private feelings, but only to *public* matters; and that so-called public objects were really abstractions of a changing universe—parts chipped out of the world for the purpose of examination.

I want you to look a little more closely at the scientific man and at his historical background. For science, as we have seen, cannot be properly understood except in relation to its background, and the material that this background presents to the scientist for study. Progress may give the appearance of steady growth, but when one looks back, at times it appears to have been made by jumps.

Take Language. Have you ever stopped to consider what an extraordinary invention that was

—a means for *expressing* feelings, *indicating* external objects, and *explaining* an argument. It is not difficult to imagine how the ape-man learned to associate a particular danger with a particular call. Individuals with throats and lungs and nervous systems built according to the same plan might be expected to give much the same cry in an emergency. It would be a public word for a private feeling. But what genius hit upon the idea of making a particular sound stand for an *external* object? This sound means a stone, that stands for a cloud, and so on. We can, of course, imagine ways in which this might arise. For example, as soon as the hand had developed the power of gripping, so that things could be used, the sound expressing the emotion aroused by the use of the thing might well become the name of the object. One of our ancestors picks up a fallen branch and *swish . . . swish . . .* goes his *stick.* I am not, of course, suggesting that a word like *vacuum-sweeper* originated in that way! But however it happened, it was a revolutionary step in history.

For science it was important in two ways. It meant in the first place that it was going to be much easier to concentrate attention on the object in future. Give a dog no name at all and it is as good as non-existent. Give an object a name and it seems to acquire a separate existence. We can then play with it, use it, and experiment with it. It becomes an object of study. But it meant also that an object was being represented by a sound. It was the beginning of speech currency. Individuals could exchange sounds instead of the objects they represented. It was possible for one ape-man to consign another elsewhere without actually going to the trouble of taking him to the place. *It was a form of economy in action.* It was a type of symbolism, the substitution of a vocal sound for a concrete object. You will see directly that the next step in this amazing advance was a form of economy in thought that led directly to mathematics.

This was the invention of a picture or a special mark to represent an object, the sound for the object, or the idea of the object. In musical notation the mark stands merely for the sound. In pictorial art the mark stands for the object—although the private feelings of the artist may be also represented. In writing or printing the marks stand for the object and its sound. Roar, for example, not only means a noise, but sounds one. When the object is past— yesterday, for example—the word stands for the idea of yesterday. And when the mark came to be merely a symbol for the idea of the object, for the abstraction, in fact, we have the beginnings of mathematical symbolism. The idea is quite simple, and worth mastering. Take the idea of *number*, for instance. I seem to remember a picture by Heath Robinson—a cave-man sitting on the ground, gazing perplexedly at three shells he had placed side by side. It was called, if I remember correctly, "The Birth of the Idea of Three." At one side of the

picture stood his cave-wife holding new-born trip-
lets. The birth of the idea of three! What I want
to bring .out is that at some stage in the early
history of man, *number* as an idea, apart from the
objects numbered, dawned upon man. It was an
abstraction forced upon him by his experience—
three stones, three clubs, three trees. . . . It enabled
man to see beyond the individual object and to deal
with objects as classes. It was the first step in the
history of mathematics; the second step in the
history of science. It became possible to put down
a mark—a stroke, to represent one *anything*, and a
different mark—a thing like a curled-up snake, for
six. Later, very much later, came the extremely
sophisticated idea of including *nothing* or *zero* among
the numbers.

Mankind was now equipped with three essential
ideas:

(1) The power of recognizing a separate object
or an aspect of an object. This is the power
of abstraction;

(2) The idea of using a mark or symbol to represent that abstraction;

(3) The idea of making a number or a sign stand
for measure or quantity.

We can see how man has used these three ideas in
his scientific analysis of the world around him. He
has simplified the complicated things of life, and he
has dealt with aspects of them numerically. You

are yourself, for example, a whole medley of abstractions which can be isolated and measured.

Consider your speech. Your throat is a source of sounds, and for a complete understanding *sounds* must be studied—sounds isolated from you as a person. The scientist, therefore, sets up a convenient sound-producing agency—something he can change and experiment with, unhampered by a living being. Your body is a source of heat and therefore the scientist will have to study how heat is radiated or conducted away from a surface. It is better for him to experiment with a whole series of surfaces, unhampered by a particular living being. You consume food—a number of complicated chemical processes take place inside your body. The scientist finds it necessary to study these chemical changes in detail under conditions where measurements can more easily be made—outside the body, in specially constructed glass tubes.

You see what has happened. When the scientist comes across complicated things he breaks them up into a number of aspects or abstractions which can be separately measured and studied, and afterwards considered together in an attempt to explain the working of the whole. Science is a revolt against the apparent wholeness of things. It tears to pieces, and puts together again. Let me suggest the following experiment to you. Decide on some object, mention all the aspects of it that suggest themselves to you, and see whether they appear to you to be measurable. Do not restrict yourself merely to those that appeal to the eye, but also to all your other sense organs. Try the game with different kinds of objects—a chair, a dog, the fire. Consider such characteristics as size, colour, sound, taste, brightness, hunger, fear, happiness. Ask yourself what you would need in order to measure and estimate these things. Ask yourself *how* you would measure them. If you cannot find out how some of these characteristics can be measured it may be due to the fact that they are not public but private aspects of the object, and that they are not measurable at all. For measuring is a public matter.

What I am really asking you to do is to look at the world *objectively*. Here let me confess to a rather cynical but amusing game I often play. To be successful at it you must practise it. Next time someone gives you a long, involved explanation of his actions, or of what he thinks or believes or asserts, or what his motive was on such and such an occasion, try to look at him objectively as a member of the ape family making noises, pay the minimum of attention consistent with politeness to what he actually says, and the maximum of attention to the reason why he is telling you all this. You will be surprised how unconvincing his explanation sounds. The game is an attempt to take a public view of behaviour, to treat man as an object in the universe, to study him without being befogged by his private descriptions of his own conduct. It is the game of looking at man scientifically.

We see then that science—even from its very beginning—has always issued a challenge: things are not what they seem. In this way science is always apparently at loggerheads with common sense. For common sense always accepts the traditional, while science is always just fresh from new experience, and from taking things to pieces. It is for this reason that so many of the advances of science—even those which are most theoretical—can be traced directly to the material needs of man. For man must live, and his scientific theories must be the theories of practice or he will not survive. Sooner or later he will fall foul of hard fact. Progress comes from the continual clash between theory and practice—between the abstraction he makes and the reality he encounters.

In this way the traditional belief that the Earth was the centre of the universe was overthrown towards the close of the Middle Ages. Trade on both land and sea was rapidly expanding; Columbus was making his great experimental voyage to America and the need was arising for accurate knowledge of tides, of magnetic deviation, of latitude and longitude. The skies came under closer scrutiny, and the accepted scheme of the heavens, as embodied in the theology of the times, was threatened with examination. The Church threw her whole weight against the new ideas once it was realized

that her authority was threatened. I need not repeat
the well-known stories of the trial of Galileo and
the sufferings of Kepler. They are worth reading
and are told in most histories of science. These
men fought the battle against established tradition
which has gone on throughout the ages.

The scientist, whatever may be his individual
character, belongs to a class that openly avows a
revolutionary policy, a class which challenges the
accepted, the complacent, the uncritical and the
doctrinaire. Science will accept nothing that rests
merely on authority. In matters of belief about the
physical universe science is an ever-present challenge
to authority; and the scientist turns to the world
about him to justify his challenge. The discovery
of truth is not in itself a disturbing event. What is
upsetting is the possibility that if the new truth is
admitted and the knowledge acted upon, drastic
changes may have to be made in the comfort and
well-being of certain groups of individuals. When
it became clear, after a direct appeal to the world
of actuality, that men were no longer prepared to
believe that the Earth was the centre of the universe,
the authority of religious orthodoxy was directly
challenged, and a revolutionary utterance had to
be stifled at birth. Three hundred years later established authority was still sufficiently strong to put
up a strenuous fight against Darwin's theory of
man's close affinity with all the other beasts of the
field.

The fight has always gone on, nor is the reason difficult to discover. Society has developed two kinds of institution. There are scientific institutions which exist for the discovery of natural knowledge —research laboratories and learned societies. There is no authority vested in their hands other than the right to pursue their investigations in so far as funds permit. Beyond this they need do nothing, and society expects nothing of them. Then there are those institutions, such as the Church and the Law, whose function it is not to extend, but to preserve existing knowledge. Since the object of all scientific institutions is to make discoveries which must affect belief, social institutions, hindered by vested interests which have grown up around them, must always lag behind the newer knowledge and offer resistance to its acceptance. The history of science provides the evidence that this statement is no theory. Science by its very nature challenges the idea of *fixity of belief*, and in virtue of that very fact is always in revolt. But in the last resort the force of actual demonstration has always triumphed against the forces of ignorance, however well entrenched they may have been.

It is hardly necessary to remind you that science is not without error. Scientists are themselves part of that great procession that stretches from ape through Adam to Einstein, and scientific standards now considered essential were unknown in the past. Moreover, it is not difficult to find illustrations of the way in which modern standards are ignored. One hears of discussions on *The Evolution of the Universe* at scientific gatherings, in which lack of precise information is compensated for by a profusion of speculation, and in which an objective observer would be surprised to hear that the evidence for life on other planets was "personal feeling" and the future of the universe was conditioned by "personal beliefs" about survival after death.

The history of science reveals four outstanding challenges to accepted belief. In each, science has given visual proofs that beliefs held without evidence, and accepted merely on authority, have been false. First there was the movement that deposed the Earth from its accepted position at the centre of the universe. That was the first great blow at man's egoism. Second, there was the great generalization associated with the name of Darwin, a theory which gave a great shock to all those who believed in special creation. Nowadays we are accustomed to evolutionary ideas and accept our position among the mammals as a matter of course. It is not easy to remember that the story of Adam and Eve and the Serpent was widely thought to be an historical account of the creation and fall of man. Much of the fabric of orthodox religious belief was woven around the story. A feeling of uneasiness began to make itself felt among honest-minded religious people when it became apparent that the fossil record pointed to a different interpretation. Some, of course, refused to consider the evidence of fossils—Hugh Miller, the Scots geologist, is said to have insisted that the fossils had been placed specially in position by God to test men and their allegiance to Him. One cannot help sympathizing with those who, having no conception of the evolutionary nature of knowledge and truth, and taking their stand on the literal truth of Genesis, found their whole position undermined. It was not surprising that they turned their faces resolutely from the newer ideas.

The stubborn ignorance of established institutions can perhaps best be realized from the well-known passage at arms at the British Association in 1860 between Bishop Wilberforce and Huxley, the great protagonist of Evolution. Bishop Wilberforce, facing Huxley with a smiling insolence, begged to know was it through his grandfather or his grandmother that he claimed descent from a monkey? Huxley turned to his neighbour and said, "The Lord hath delivered him into my hands," and rising to his feet he made his memorable reply, "I have certainly said that a man has no reason to be ashamed to have an ape for his forefather. If there were an ancestor whom I should feel shame in recalling, it would rather be a man of restless and versatile intellect who plunges into scientific questions with which he has no acquaintance, only to obscure them by aimless rhetoric and skilled appeals to prejudice." Of course the effect of the discussion was to give the new viewpoint a publicity it would not otherwise have received.

These two revolts, led by Copernicus and Darwin, paved the way for others. The third challenge was to the belief that the matter of which the body is composed and the processes that go on in the body are necessarily fundamentally different in kind from those that proceed in the ordinary material world around us. This was the next step in establishing the uniformity of Nature. And lastly there is to-day the movement which seeks to break definitely with the world of restricted common sense and offers us a universe finite in extent, continually expanding, incapable of detailed description in terms of common experience, and only to be described by means of abstruse mathematical symbols.

It is the driving force of science that brings it into revolt in the manner I have shown. This does not mean that individual scientists are themselves rebels against prejudice. On the contrary, in common with the rest of mankind, they have their prejudices, they are biased, and endeavour to explain away anything that makes their own accepted beliefs uncomfortable. No one has adequate evidence for ninety per cent of the beliefs he holds, and individual scientists are not different in this respect from the rest of mankind.

3. SCIENCE IN ACTION

LET us now turn to some of the material of science in order to study the methods which have been evolved for its examination. The piece of the universe I propose to ask you to examine is something which has never been seen but which everyone has felt—the *air*. We have felt it strike our faces, it has chilled us to the bone, we have heard it howl and shriek, we have smelt it—but we have never seen it. And yet, unseen as it has been, "blowing where it listeth," we have mastered it and turned it to our own use. How has this been done?

The range of motions and of pressures of the air is enormous. Place your face close to a loud-speaker or to earphones and you can detect no air disturbance; and yet the very fact that you can hear sounds shows that the air actually is in gentle motion. You wave your hand close to your face and you can feel a gentle breeze. Outside in the street a piece of paper is being blown to the housetops, the wind pressure upon it is as great as its weight. You must have seen a hoarding or a deeply rooted tree levelled by the sheer force of the wind. The pilot of an aeroplane travelling at 400 miles per hour would probably break his arm if he suddenly exposed it to the direct blast of the wind. It is this capacity for exercising pressure which air in motion possesses that scientists have turned to advantage.

What else do we know about air? It has weight, of course, which can easily be verified by weighing a flask full of air and then when pumped empty. What else? Well, try this experiment. Light a cigarette. Draw in a mouthful of smoke. Bring your mouth close to a polished surface—say a sheet of glass, a polished surface of table, or a sheet of paper. Now breathe the smoke *gently* on to the surface, without too much draught. Look how the smoke seems to cling to the surface. Now blow very gently along the surface, and notice how little waves appear on the surface of the layer of smoke, rapidly rolling up into a sort of eddying motion, and breaking up almost like spray. The smoke still hangs on close to the surface. It seems to be sticky. Is it just the smoke that is sticky or is air also sticky? How can we find out?

How would you find out if a *liquid* were sticky? You would dip something into it, say your hand, and see if the liquid stuck to the surface. Make the experiment. Put a coin in a bowl of water, rub the surface of the coin with the water while it is immersed and then *pull it up* into the air. When you return it to the water you will find that the coin is covered with bubbles where the air has stuck to its surface. Air *is* sticky.

So air is heavy, it is sticky, and it can exert a fairly strong pressure against a surface on which it beats *directly*.

Have you ever noticed that if you are standing

in a wind and want to listen carefully you turn your head round so that the wind does not beat directly on to the ear, but moves smoothly *past* it? What happens to this pressure when the wind does not meet the surface dead on but runs across it?

Make this experiment. Take a flat sheet of paper and hold it with both hands, by a thumb and forefinger at two adjacent corners. The other two corners will droop away from you. Now blow steadily and strongly across the *top* of the paper from the front to back. Although you are blowing along the top and not on the *under* surface of the paper, you will find the paper gradually rising until it is almost horizontal, as if it were being forced up from underneath or sucked up from above. We conclude that this pressure exerted by moving air does not merely offer resistance to movement straight through it, but if the surface is placed edgeways to the wind, or if the surface moves edgeways *through* the wind, a lifting force upwards is produced on that surface. This is the principle of the aeroplane wing. If the wing can only be driven forward, it will at the same time be sucked upwards. Broadly speaking, to solve the problem of aeroplane flight we have to ask how it is possible to get the greatest lifting effect from the wind with the least expenditure of effort in driving the wing forward against the opposition of the wind.

Consider an aeroplane. Strip from it all the things that do not matter for our purpose. Strip away the wheels, the under-carriage, the cabin; eliminate the tail and the fuselage, everything, in fact, except the wing surface and the propeller. We ought to keep the engine, because the propeller must be driven, but if the presence of the engine makes it too complicated, throw the engine also overboard. All we want is a wing surface and a rotating propeller. The propeller is so shaped that as it is spun it acts like a fan, buffeting the air, and grabbing it to itself and discharging it behind in a steady stream. As it pushes the air away from itself backwards the propeller is at the same time pushed forward. Have you ever stood on ice and tried to push something away from you, and found that you were also pushing yourself back at the same time? We have a similar situation here. The principle is a general law in mechanics—action and reaction are equal and opposite. When I was a boy we used to make a steamboat by getting an egg, which had been emptied through a small hole at one end, filling it with water, resting it on four vertical nails on the boat, and boiling the water in the egg by means of a candle. As the steam was driven out through the hole, the back pressure, or the reaction of the steam on the air, pushed away the boat in the opposite direction. The same principle was at work. In the aeroplane the propeller—carrying with it, of course, the rest of the aeroplane—pushes itself forward by kicking against the air. As the wing is thus driven forward, the suction effect on the upper surface of the wing comes into play; a great draught is created by its own motion forward, and the whole machine is *sucked up*. The machine is in flight.

Of course there are all sorts of complications in practice. You cannot keep the propeller turning without a fairly heavy engine which has to be lifted. You cannot lift such a heavy mass of material and send it along at high speeds and expect it to hold together with string. A great deal of weighty structure is needed to hold it together, and that extra weight has also to be lifted. Moreover, as the aeroplane is rushing through the air at great speeds, these extra parts are offering resistance to the air —they act as an obstruction. And in order to drive against this you require stronger and heavier engines. There are a whole series of further complications, but I shall limit our discussion to a consideration of air pressures only.

There are two problems of importance that call for solution:

(1) How should the exposed parts of the machine be shaped, in order that when these parts meet the onrushing wind during flight the drag or resistance is as small as possible? To answer this question we must study the way in which resistance depends on shape.

(2) What lifting force and what resistance is experienced when a wing of a particular shape is moved through the air at some definite speed? We must study the relation of lifting force to speed.

Let us see how the scientific man attempts to provide an answer to these two questions.

For this purpose I want you to go with me to an Aerodynamics Laboratory. We enter a large hall down the centre of which runs a single piece of apparatus, in appearance like a long wooden box, seven feet square and about forty feet in length, rigidly fixed to the ground on metal supports so that it stands about ten feet above the ground. This long object is not really a box, for it is open at both ends, so that it is more like a long horizontal wooden tunnel suspended in mid-air. It is approached from the ground about half-way along its length by a flight of wooden steps leading to a glass door in the side of the tunnel. Let us mount the steps, open the glass door, and step into the apparatus. We find ourselves in an elongated space seven feet high, seven feet wide, and about forty feet long, opening into the outer room or hall at both ends. Not, however, quite freely, for at one end the passage is blocked by a propeller. Through the centre of the floor of the tunnel there projects vertically a metal rod, fixed to the top of which is a beautiful little model of an aeroplane, complete with wings, struts, body, airscrew, tail, rudder and fins, and so on. As we look we hear a click from below and the propeller at the end of the tunnel begins to turn, slowly at first, then faster and faster, sucking the air through the tunnel in a steady stream. It soon becomes clear that if we wait much longer we shall find ourselves in a terrific gale. We are out just in time, for, with a roar, the wind is tearing along the channel at a speed of nearly sixty miles an hour. From the safety of the steps we look inside through the closed glass door; the little model aeroplane, perched securely on its vertical rod, is swaying very slightly—almost imperceptibly. Nothing more seems to happen, so we descend the steps—the roar of the wind is beginning to make our ears throb. Below we find two people at work, oblivious to the intolerable noise. One is seated just below the spot where the model aeroplane rests in the tunnel. He is busy placing delicate weights on to a horizontal metal beam, attached to the rod which goes through the floor of the tunnel to the model above. We discover that he is balancing with weights the wind forces on the model, so that he can tell with great accuracy how much lift and how much resistance the model machine experiences in a known wind speed. He is, in fact, weighing the forces on this small model aeroplane. After each set of weighings he signals to his companion, and the roar of the wind in the tunnel increases. He then adjusts his weights for greater and greater wind speeds. Turning to the other man, we find that his job is to keep the wind speed steady during each set of weighings and to know precisely what speed it is. How does he do this? His eye is glued to a microscope focused on a bubble inside a glass tube—something like a spirit-level. A rubber tube is attached to one end of the glass tube and is connected to a hole in the side of the tunnel past which the air inside streams. He is measuring the speed of the wind in the tunnel.

But, you may object, the model aeroplane is not in flight but at rest, and the air rushed past it. Surely that is very different from what happens in practice, when the machine moves and the air is at rest. But is there any difference? Have you ever seen an aeroplane flying against a heavy wind and making scarcely any progress? It is flying nevertheless, and if the pilot were out of sight of the ground he would be quite unaware of the wind. Anyone who has been in an aeroplane knows that there is little sensation of speed. It is the *relative* speed that matters.

So far we have only seen measurements being made on a model, perhaps two feet wide, about one-twentieth the scale of the full-size machine. The force on the full-size machine is required. How can this be obtained from the measurements we have seen being recorded? Another question might also legitimately be raised. The greatest wind speed being used in the wind tunnel was about sixty miles per hour, while aeroplanes fly at speeds as high as four hundred miles per hour. How can the lifting force at the higher speed be found from measurements taken at lower speeds?

These are not easy questions to answer in a non-technical manner, but I can indicate the direction in which answers are to be sought.

Look over the shoulder of the man who is balancing the forces on the model. He has written down in one column the numbers showing the forces, and opposite each figure in another column the wind speed. We notice a peculiar thing. When the wind speed was doubled, the force went up four times. When the wind speed was trebled, the force was nearly nine times as great. When the wind speed was quadrupled, the force was nearly four times four—that is, sixteen times the original value. There seems to be a definite rule by which one could predict the forces for very much higher speeds than those measured, provided, of course, that the law continued to apply. It is important to notice this qualification—if the law continued to apply.

A similar argument applies to the question of size. It is found that if a model of twice the size, but otherwise the same shape, is put in the wind tunnel, the forces on it are also four times the previous values. If the model is increased to three times the size, the forces are increased to nine times, and so on. Again, the same type of law appears, giving us the possibility of prediction. It is by such investigations, together with numerous experimental checks on full-size machines, that the scientific man is able to use his knowledge of the model to estimate the behaviour of the full machine. In the same way the designer can make his estimates and plans with assurance.

Let us continue our journey through the laboratory, leaving the wind tunnel and passing to other parts of the large room. On one side is a collection of models that have been used in the work, parts of machines, model wheels, tail planes, wings of innumerable shapes, propellers of all sizes and sections, engine radiators. Here is a peculiar piece of apparatus designed to discover why in certain circumstances a tail or a wing will begin to flutter and possibly crumple up when in flight. There are complete models set up to investigate the forces that arise during spinning. Every aspect of the problem has apparently been thought out. Here are two investigators absorbed in the peculiar problem of *predicting* how the R 101 behaved just before her collapse. They are working from a knowledge of the forces she was likely to experience in those last few dramatic moments of her flight. In another corner stands a long wooden trough along which water is slowly and steadily flowing. Upright in the centre of the trough is a model of an aeroplane part—a strut. Thin streams of coloured liquid curl and swirl around and past the strut, forming a broad eddying wake of turbulent and disturbed water in its rear. The same kind of thing happens in air. The strut is removed and one of finer shape is put in its place. The band of eddying fluid is much reduced and less energy is lost in useless disturbance. In this way information is gradually forthcoming concerning the best shape of the exposed parts of an aeroplane, so that the resistance encountered may be reduced to the minimum. In another part of the laboratory there is a model of a wind tunnel, where experiments are being made in order to change and improve the huge instrument itself.

Let us go into one of the workshops. Fine instrument-making is here the order of the day—not the standard orthodox instruments offered for sale by scientific instrument-makers, but instruments in the experimental stage. Here are mechanics and pattern-makers who must work with extreme fineness, for failure or inaccuracy at this stage carries mistakes into all the later stages.

It is clear that scientific research has a complicated organization. Any piece of work has a number of linkages—we might call them horizontal and vertical linkages. Horizontally, scientists in the most diverse fields are dependent on the results of each other's labour. In one laboratory we have seen those engaged in aerodynamic research requiring, for example, the results of optical research to enable them to use reading instruments, requiring the work of chemists to provide the gases, dyes, and oils most suitable for rendering the flow of the air and the water visible, requiring physicists to discover the effect of changes in atmospheric pressure and temperature during the course of an experiment, mathematicians to interpret the results found, and so on. These and many others form the broad horizontal band of research workers, whose constant collaboration is essential for the success of any piece of specialist investigation.

There is a vertical band of linkages, equally indispensable. At one end of it there are accurate workers in metal and wood, who produce the delicate machinery required in the investigations; at the other those whose job it is to direct and co-ordinate the work as a whole. There are all grades of workers in this vertical column—from mechanics to administrators—but each grade is essential to the smooth working of the whole. Science is a social activity, and by this I mean not only that science relies on the community for its support, not only that the results of its investigations have important social consequences, but also that scientists are a solid and closely interlocked group, whose object is the pursuit of assured knowledge.

4. IS THE UNIVERSE MYSTERIOUS?

I HAVE tried to show science at work behind the scenes in order that two things might be appreciated: first, the close contact with concrete things that science persistently maintains, and secondly, that nowadays scientific work is not an individual activity but a corporate undertaking. It is carried on by large groups of workers dependent one upon the other for help, understanding, and verification —public verification I called it, in order to distinguish it from personal or private belief.

I propose now to discuss a much more difficult question—one which has become prominent because of the efforts which are being made by some scientific men to interpret scientific discovery and to indicate to the man in the street the direction in which science is moving. Any attempt to do this is very desirable; for the day is long past when understanding of a powerful activity like science can remain the private possession of a few. But the manner in which it has been done has revealed what a difficult, and even dangerous, undertaking it is. As I understand it, the function of a scientific expositor is, first of all, to reveal to his hearers the field within which science can operate, and secondly to interpret the facts of science within that field. A scientist, however, is also a private individual, and unless he has clearly delimited his public responsibilities, he is likely to be found gaily trespassing in a private region where science does not yet operate, and holding forth there about the things he "believes."

I cannot illustrate the dangers more clearly than by quoting from one of my correspondents. "Would you kindly inform me," one writer asks, "if you consider the statement 'there is no life after death' a scientific one, when you take into consideration the altered view scientists now take with regard to the atom." Nor is this the only occasion on which the same issue of the atom in relation to the "spirit" of man has been raised.

Let me say at once, lest there be any misunderstanding, that I would never say that there is no life after death. Having been unable, after much seeking, to obtain any satisfactory knowledge which would even enable me to state in scientific terms what the phrase "life after death" means, I cannot yet say anything positive about it from a scientific point of view. My correspondent, however, goes further than I do. His question suggests that modern researches into the constitution of the atom do provide some information about "life after death." I confess that I am completely baffled by the apparent relationship between atoms and immortality, and I doubt if the most sturdy believer in psychic research would countenance the suggestion of any relationship. And yet the belief is fairly widespread that in some peculiar way recent investi-

gations into atoms and electrons point to the discovery of something "spiritual" at the core of
scientific theory.

In order to clear the issue let me first draw a
distinction between two points. I want to contrast
two questions: *"Is Science Mysterious?"* and *"Is the
Universe Mysterious?"* It is the latter question that
appears as my title, but actually the question one
invariably hears discussed is *"Is SCIENCE Mysterious?"*
I want to show that the answer to this question is
definitely "No." There is no mystery in science.
You can make what answer you like to the other
question, for it does not come within my province.
If you ask me as a private individual what I feel
about it, of course I will tell you, but I should have
to speak privately and my statement would merely
have the standing of that of any other individual
man or woman. It would not have the backing of
the scientific movement.

"Is SCIENCE Mysterious?" is the question at issue.

One often hears it said by those who, having had
no chance of getting their science at first-hand,
have had to rely on interpreted information, that
the world is not deterministic like the inevitable
workings of a machine. They declare that the closer
scientists penetrate into the workings of Nature,
the clearer it is becoming that in the last resort
things happen of their own volition—which is, of
course, only another way of saying for no apparent
reason. They say that this indeterminism or "free-

will" is to be seen in the minute details of the structure of the universe—in the behaviour of the electron, and that it thus follows that in large-scale events old-fashioned determinism is dead, and that human free-will is an established fact.

First let me say something about the historical setting of this theme. Of course the Greeks argued about determinism, but its modern phase begins about the time of Newton in the seventeenth century. More than any other man Newton organized into definite scientific laws our knowledge of certain aspects of Nature—in particular, facts about the solar system. He showed that the movements of planets in the remote past fitted in with his generalizations, and that his laws could be used for purposes of prediction. Throughout the eighteenth and nineteenth centuries the study of mechanics—with which Newton had been specially concerned—passed from success to success, and the application of the laws of mechanics to engineering practice was followed by remarkable achievements. Newton—who, by the way, was a very religious man—thus unwittingly laid the foundations of a form of mechanical materialism which reached its zenith in the nineteenth century. The world came to be regarded as a vast complicated piece of machinery—a clockwork mechanism which, having once been set in motion, with all its intricate interlocking detail, ran smoothly and relentlessly along its predestined course, according to the laws laid down by Newton.

Much has happened since that time to alter the setting of this picture, but from what I have already stated you will recognize that this mechanistic explanation was a highly public view of the world, a view which in actual fact has had tremendous success in explaining and describing the visible behaviour of material things, not only on our little earth but throughout the solar system. Naturally, side by side with this outlook there existed, and there still exists, that private view of human behaviour which says "I am master of my destiny, I can do this or that just as I wish; my will is free." It was natural that all those institutions and opinions associated with the private belief in free-will should feel that this powerful and successful piece of scientific machinery might at any moment try to explain away personal conduct as it had explained the machine. It is possible that they felt that any such attempt would be a threat to personal happiness. Therein arises the traditional antagonism between science and religion.

Now, certain things have happened in science to change this picture slightly. I refer to the theory of relativity and to the quantum theory. Not that I propose to explain either relativity or the quantum theory, but I shall try to show how these new theories affect the problem of determinism.

It would be well, first of all, to explain in greater detail what scientists mean by a "scientific law" and by "observed facts." Everyone, given the right kind of apparatus and the right kind of skill, can verify "facts" for himself. Facts are the same for everybody. A "scientific law" is a general statement which covers and unifies observed facts. Let us take a simple example to show how the scientist arrives at his scientific laws.

You know those "Try your Strength" machines that one finds in fairs and recreation grounds and on piers—in fact, anywhere where there is time and money to waste. One is given a hammer and is told to strike upon a block; the strength of one's blow can be seen by the rise of a ball which travels up a column. If your strength is as the strength of ten a bell rings, otherwise you fail to get your money back. The principle of the thing is that the strength of your blow is measured by the height to which the indicator rises. Suppose you strike the block with a certain known force and watch how high the indicator rises. Then if you give the block a blow twice the strength, you will find the indicator rises to about four times the previous height. Use three times the strength and the indicator rises to about nine times the height, and with four times the strength the indicator reaches sixteen times its first height.

On seeing these figures you say at once that there seems to be a rule or law which relates the strength of the blow and the height to which the indicator rises. I will state the law:

The strength with which the block is struck is propor-

tional to the square of the height to which the indicator rises.

I want you to notice two things.

First. We seem to be in a position to state that if we hammer with five times the strength the indicator will rise to five times five, or twenty-five times the height to which it rose on the first occasion. That seems obvious from the results already found. But I would remind you that there is no experimental evidence for it. In order to be absolutely sure we should have to strike the block with five times the original strength and find out the height to which the indicator rose. It is true that we are induced by the results we have already obtained to make a scientific prediction about the behaviour of the indicator. A scientist would say that an *induction* had been made—i.e. that from a series of observed facts a wider law uniting them all had been formulated. Notice that in making our induction we have overstepped the experimental evidence. Far from being a definite *assurance* about what will happen in the future, scientific prediction is merely a statement that under such and such definite conditions a certain result may be expected. In other words that expected result would be consistent with our generalization. Let me state this first observation categorically:

A scientific law is only a statement of what seems extremely likely to happen.

Scientific law does not deny that further facts

may still come to light which are not in accordance with that law.

Second. It is often said that one of the objects of science is to state in advance what will be found in the future under given conditions—in other words, to predict future experiences. What do we mean by experience here? Consider the "Try your Strength" machine again. A blow had to be struck in order to observe the height to which the indicator rose. Now when we double the striking force we find that the height is not exactly quadrupled but only very nearly quadrupled—sometimes the height is a tiny bit more, sometimes a trifle less. You might say this was an accident. Try again—and let us suppose the machine is as perfect as science can make it and that the required striking force can be exactly produced. On ten separate occasions suppose the block be struck with exactly twice the strength. On no occasion does it rise exactly to four times the height —only very nearly so. You will probably cast about in your minds for the cause of this apparent error. You say that it is so small that your observation must be at fault; that the ball which acts as an indicator "intends" to rise to exactly four times the height on each occasion, but for one reason and another it fails to do so, and so on. But if you strike the block with twice the force a hundred times, or a thousand times, instead of ten times, you will still find the same small error. You will then turn your attention to an examination of the law, and you might come to the conclusion that the law as you had framed it was *too perfect*. You might say that a law to be a perfect law ought to embrace these minor errors of observation. You might say that you do not want an idealized law but that you would prefer a working law, one which would give you the odds on getting one of a series of measurements, all, of course, very nearly equal. The fact that you have got an idealized law results from the method of abstraction which I have already discussed. By the method of abstraction you have stripped off from the ball and from the blow all the things you imagined to be of no consequence in practice. You have got an idealized, or perfect, law. It may not, and generally does not, meet the observed facts of your real ball and your real blow.

You will see from all this that science makes idealized laws and uses them for predicting what is likely to happen in the future. These future events may not be *exactly* as they are forecasted, but they should nevertheless approximate to the prediction.

So if we look back on the rigid determinism of the nineteenth century we see that it was too hard and fast. This rigid determinism took scientific laws as they were set out, sharply and clearly, by Newton and his successors and assumed that they were the laws of Nature. The assumption was made that if only one measured *accurately* enough one would find these laws *exactly* fulfilled. In other words, the determinism of the nineteenth century considered that these exact scientific laws were really Nature's laws, and that what you actually found in the world of reality was only an approximation to those perfect laws. We now see, however, that this was putting the cart before the horse. Scientific laws no longer occupy the magnificent and impregnable position they once did. Whatever validity a scientific law has is shown by the fact that the law is a good approximation to the operations that actually go on in this complex and changing world. Scientists repudiate the idea that it is possible to predict *absolutely* the features of the universe.

We are now ready to turn to recent researches connected with the electron and the quantum theory.

The electron is a scientific abstraction which is causing people a great deal of trouble. If only one could get hold of a *single* electron and make experiments on it one might manage to make short work of it, but electrons exist in groups and thus render themselves immune from too close a scrutiny. One has to deal, therefore, with the average of the group—the typical electron. Now electrons are quite unlike anything that science has ever experienced in the whole course of her history, but electrons appear to be a basic constituent in matter, and so any discussion of matter at that level involves an explanation in terms of electrons. Hence our difficulties.

First of all let me remind you of the relation

between atoms and electrons. For most purposes a lump of matter—a piece of iron, for example—behaves as a whole and remains one piece unless you subdivide it. In describing the behaviour of this piece of iron Newton's laws of mechanics approximately hold good. Now in theory our piece of iron can be regarded as consisting of an enormous collection of atoms. During any chemical change these atoms remain intact. Once an atom, always an atom, so to speak. But if you pour some acid over the lump of iron it may dissolve. The lump of iron as a whole will have disappeared, but the atoms have not disappeared; they can all be accounted for. They will individually have united with the atoms in the acid. The presumed behaviour of individual atoms in a lump is only part of the story, however. For an atom, as you know, is supposed to consist of a charge of positive electricity round which there circulate at incredible speeds, and at various distances from the central positive charge, one or more tiny electrons—like planets circulating round the sun. That is the theory. These whirling electrons are thought to be charges of negative electricity. From substances like radium they are shot out naturally, and in the laboratory they can be knocked out of the atom if you hit it hard enough with certain rays—metaphorically speaking, of course. Now the electrons move around the central charge at various distances from it, but on occasion they seem able to pass from one path to another path. We only deduce that they have changed paths because they have radiated light in doing so.

It is necessary to digress for a moment in order to comment on the precision of our knowledge in these fields. Ultimately, of course, all science depends for its accuracy on the precision of experimental measurement. When we come to deal with the ultimate particles we call electrons and try to measure distances and times—distances between two neighbouring electrons and times of passing from one path to another—we are working at the very limits of scientific experience. The very act of measurement affects the thing we are trying to measure, for our effective measuring-rods and the objects of our measurement are both ultimate particles. There is thus a fundamental difficulty of measurement in these fields. Suppose, for example, you tried to find out how cold some object was by touching it with your hand, and suppose that every time you reached out to touch it the heat from your hand caused it to melt. The same sort of situation is created in attempts to measure the electron. With the ordinary conventions of space and time used for measurements and predictions of speeds and distances of largish objects, an odd thing happens when the tiny electron is measured. Its speed and its position cannot be measured independently. The more accurately you fix the speed the less accurately can you fix the position, and the more accurately you can measure the position the less accurately can you measure the speed. Speed and position are not independent and separate aspects of a moving electron. If you wanted to predict where a particle was going to be at some definite instant in the future you would, of course, require to know where the particle is now, and at what speed it is passing through its present position. It is just the same for a train. If you want to know when the train from A will arrive in B you must know the time it started from A and the speed at which it is travelling towards B. But unfortunately these two things cannot be found separately for the electron, so that you cannot describe in detail its behaviour as if it were an ordinary particle.

The reason for this difficulty is known. It resides in the fact that there is a minimum quantity of energy—the quantum it is called—which is capable of taking part in any action. This quantum acts as a whole. There is no half-way house. A quantity of energy is either a quantum (or several quanta) or nothing. You cannot have a fraction of a quantum. As the electron moves on its journey and gets from one path to another it gives out a quantum of energy which shows itself in the form of light. Now this quantum of light does two things. It enables us to detect the presence of the electron, but at the same time in being emitted it jerks the electron right off its balance, so that we do not see the electron where it *is* but where it *was*. In this way it becomes impossible at the same time to fix both the position of the electron and its speed.

There is no mystery about the idea of the quantum, yet some writers have suggested in popular expositions that here is a problem that by its very nature eludes determination and so is a mystery. But I would ask you to remember this—these writers do not suggest that the electrons do not go along paths of *some* kind. They are prepared to state what proportion of a group is distributed along one path and what proportion along another, but they say that it is impossible to predict for an individual electron which path it will traverse. They can state how probable it is that it will go along this or the other path. So there is really nothing here to create confusion. Since it is known how the difficulty has arisen there is nothing to support the feeling that mystery lies at the very heart of the universe. After all, every scientific law ought to be stated only as a *probability*, and any prediction, as we have seen, is only a statement about what will probably happen.

The point I want to make is this. The newer physics in its study of the electron has merely made us recognize that there are limitations to the field in which man can make accurate predictions. This is not a new conception; it has been recognized in other spheres. Man is limited, for example, by the very nature of his sense organs. He is limited by his heredity and by his environment and by his social tradition. Even scientific truth is a limited and temporary statement depending on the state of knowledge of the time. It is true that a new limitation has been exposed by this work on the electron, but there is no *mystery* about it. It has not affected the determinism which was essential to scientific method before these recent developments in physics. Predictions which were made—and are still being made—on large-scale objects remain valid.

Finally, remember that what we have been discussing has nothing whatever to do with what is called the "free-will" of human beings. Free-will is a purely private belief, a purely personal interpretation of human conduct. Science does not take account of such beliefs. Next time you hear it suggested that modern physics has knocked determinism out of science, I hope you will call to mind the considerations I have here put before you. Without determinism there could be no science.

5. SCIENCE—DISRUPTIVE AND CONSTRUCTIVE

Look round the room in which you are seated and draw up a list of the things you can see in their natural state—just as they are found before they have been tampered with by man. There is coal on the fire, water in the glass, and possibly flowers in the vase. There seems to be little else. Among all this collection of things—chairs, cutlery, books, wall-paper, clothing, curtains, electric switches, carpet, pictures, nails—what is there true to nature? Even the water might be soda water, the fire an electric one, and the flowers imitation.

The difficulty you have in finding "raw material" in your room will impress on you the extent to which modern civilized life has come to depend for its necessities and luxuries on the refining processes of industry. Here are two pieces of paper. One is coarse but strong. I find it difficult to tear. It will not lie flat. I hold it up to the light and I find that there are uneven clots in it. The other piece is smooth and thin. When it is held up to the light its texture is uniform. It lies flat and is suitable for writing on. Think of the amount of labour, skill, and research that have gone to build up these two types of paper, each suitable for a special purpose. Here is a fountain-pen with its iridium nib and its vulcanite container; there is a fire-place with its cast bars and ornamental tiles. Look at the carpet or curtain with its intricate weaving and fast dyes. All these things are of highly elaborate manufacture, dependent on the application of complex scientific processes. They are part of the ordinary furniture of our lives. In this practical guise science has insinuated itself so deeply into our homes, work, and amusements that we are as little aware of it as we are aware of our own breathing. It is not as if science were an old-established tradition. There is scarcely an electrical device in common use at the present day that is more than thirty years old.

If these scientific amenities were suddenly withdrawn, our civilized life would crumble. Life would become impossible in our cities, with their large massed populations, dependent on the smooth running of power stations for the preparation of food, and on mechanical transport for its distribution. As a community we have built on the assumption that these things will go on, and that we may depend on science somehow or other to satisfy new needs and to overcome new dangers. Try to imagine what steps you would take to-day if it became known that by to-morrow morning every scientific discovery and invention of the past century would have vanished, leaving you to cope with the demands of home and communal life. I suggest this speculation not for idle amusement but because I want you to appreciate what I mean when I say that we have staked our future and the continued functioning of our civilized life on a complete belief in science.

I should like to explain more fully some of the implications of this fact. Consider what Britain was like 150 years ago. It was a country with a much smaller population, a proportionately larger agricultural industry, a strong merchant class, and the small beginnings of industrialism. Then came the new sources of power, coal and steam, and the face of Britain rapidly changed. Cornfields and pasture lands gave way to coal-mines and iron foundries; quiet villages suddenly expanded into busy towns and the country air was transformed into the smoky atmosphere of cities. The rural population marched steadily into the rapidly growing towns. In a generation, Britain gave herself over completely to the new mechanical age. Three facts are important. First of all that the whole mode of life of the greater part of the population of this country was completely changed *in one generation*. Secondly, that it was the discovery of new uses for steam, coal, and iron which was the main factor affecting the structure of Britain's population—just as its structure had previously been largely determined by agriculture. And thirdly, that the kind of life which our grandfathers built up in the towns rested on the assumption that the new movement was permanent and could be relied upon to last.

We have lived to see how false was that last assumption. The needs of a community are not static. Urged on by these changing needs, scientific investigation is itself a continual spur towards change. Moreover, scientific progress cannot be localized: it knows no national frontiers. Mechanization spread rapidly but unevenly. Nor was it ever asked how much mechanization was necessary to supply the world's needs, or which countries were best adapted to the process. In the absence of international action acute competition in world trade resulted. Nor did Britain ever attempt to reorganize her own industrial machinery in order to maintain her early advantages. She did not even make full use of her scientific resources, which were considerable.

The immediate changes which followed the coming of the Steam Age were merely the prelude to a continuous transformation, whose driving force was the needs of the community. As science and technology satisfied each need there followed far-reaching changes of a social nature. This sequence of events will come home to you if you try to draw up a list of the industries that have come into existence—not forgetting those that have died in the meantime—during this past generation alone. Let us ask ourselves whether we or our friends could have found employment in our present trade or profession fifty years ago. Consider an industry like transport and take note of the succession of horse, train, motor-car, aeroplane. Even "brain-work" is being highly mechanized: in insurance offices and banks one hears the soft hum of the computing machine, adding and subtracting totals and sub-totals. In the scientific laboratory electrical machines perform the most complicated operations, differentiating and integrating, computing horizontally and vertically, enumerating the results, and even going to the length of recording mistakes. Not only is the machine thus taking the place of man, but it is making man's work more accurate. Steadily and relentlessly hands and brains are being replaced by machines, and the making of these new machines becomes itself a development of industry. Let us examine one such development in detail.

When coal is heated in a particular way, three main products are separated out—coal-gas, coal-tar, and ammoniacal liquor. As early as 1789 the manufacture of coal-gas was begun; fifteen years later a public gas-works was established. Gas-lighting spread. It was a social asset. The by-products of gas-manufacture, however, were a social evil, and became a nightmare to works' managers. Lighting was a social necessity, but it was equally a social necessity to get rid of its evil-smelling consequences. It was the scientist in the end, not the commercialist, who turned to a study of these waste products. Bethell discovered that one of the oils from the waste coal-tar was an excellent wood-preservative, a discovery that led to a new industrial process and incidentally solved the nation's problem of rotting railway sleepers. In 1825, Faraday isolated benzine in the laboratory. Twenty years later benzine was discovered in coal-tar—coal-tar which had for years been thrown away by enterprising business men who had even paid large sums to have it removed. To pay to have it examined by a chemist was a notion which had not presented itself to them. In 1856, Perkin, a young chemist aged eighteen, while experimenting with aniline, a substance obtained from the benzine in coal-tar discovered the first aniline dye. This dye was *mauve*, and its discovery marked the development of the modern dye industry with all its subsidiary interests.

The industries which have directly followed on a scientific study of these waste treasures of the coal industry are almost legion. Dyes of all shades, naphthalene, ammonia, creosote, benzol, pitch, tar, perfumes, paraffins, drugs, flavourings, disinfectants —to say nothing of high explosives. Industries concerned with paint, india-rubber, varnish and stain, composite fuel, wood preservation, felt manufacture, to name only a few concerns, depend on the salvage work of scientists who have rescued valuable raw materials from waste rubbish.

The moral is clear. The present and future needs of a community cannot be provided for adequately and intelligently unless appropriate machinery exists for the application and development of scientific principles in industry. From the birth of the scientific idea to the manufacture and sale of the finished product there must be co-ordination.

A significant test of intelligent planning in industry can be made by examining the effectiveness of this machinery for adjusting supply to present and future needs, and of exploiting the natural resources of the country. In the application of science to industry there are three important stages. First of all the discovery in pure science has to be made. There are the physicists, chemists, biologists, biochemists, bacteriologists in their respective laboratories concerned mainly with fundamental theory, and the opening up of new fields of inquiry. Their work is carried out principally in the universities, in Government scientific institutions, and in specialist museums and libraries. The results of their investigations are to be found in the journals of the learned societies—freely given to the world and often published at the expense of the investigator. These scientists struggle along, begging for grants towards the cost of apparatus and publication; and, in the past, help has been given grudgingly, as a luxury which the community could ill afford—although the history of technology has shown that scientific research is a vital necessity if communal life is to be developed successfully.

The next stage in the process brings us to the factory laboratory, where the theoretical idea of pure science has to receive its industrial baptism. The function of the industrial laboratory is best illustrated by an example. Take dyes. A new dye has been isolated from coal-tar by the chemist. How fast is it? How does the crude dye require to be treated to make it commercially useful? How much useful dye is obtained from a given quantity of the coal-tar? What is it going to cost to extract it? What sort of commercial plant is most suitable? How are the various stages in the extraction to be arranged in the factory? What other materials not immediately available will be needed? How reliable is the supply? What will the raw materials cost? Are there any other processes that might be handled with advantage at the same time? Are there any by-products and what is to be done with them? There are many questions which call for careful study and experiment with model plant. At this stage these questions cannot be answered either by the pure scientist or by the manufacturer. They require technical skill and an understanding of full-scale factory conditions, as well as of scientific experimentation.

An institution for the study of such problems I call an industrial laboratory. It is an essential development, and an industrial system that does not deliberately provide for it can never hope to be efficient. The stream of pure science will pass it by and the system will become obsolete. It is not enough to have one or two men with doubtful scientific qualifications, underpaid and overworked, to handle the transference. The link has to be a strong one, or subsidiary industries will not develop from the main stem as they must do if the changing needs of the community are to be supplied. In this country industrialists as a class have failed lamentably to recognize this need. Among the multitude of industrial activities on whose well-being the country rests there are very few indeed in which this aspect of industrial planning has shown itself. For want of an effective channel of communication, a mass of pure scientific knowledge probably lies industrially sterile in the archives of the learned societies.

Every industrialist, every manufacturer, every business man who has entrusted to himself the task of carrying on industrial development, can now ask himself whether during times of peace and plenty he took serious thought of how new scientific knowledge was to be utilized in his own manufacturing processes. Has he ever considered what experimental processes would have to be devised and tested in order that the natural resources of the country could be explored and used to the fullest extent? By natural resources, I refer not only to material and plant, but to scientific knowledge.

One does not have to wait for an answer to these questions. The answer is known. If politicians and captains of industry in whose hands the security of millions of our workers rests had been alive to these problems, if they had been able to see production and consumption as a scientific whole, the disastrous gap between pure science and practical manufacturing processes might long ago have been bridged. New ideas not merely make existing industries more efficient, but they create *new* industries—and this is infinitely more important. The development of new industries, based on a full knowledge of natural resources and an economic scientific co-operation with other countries, could build a new industrial Britain.

It is worth while considering some of the reasons why there are so few developments of the kind I have indicated. It may mean that our captains of industry have not been alive to the importance of scientific research and to the necessity of utilizing it as soon as it becomes available and arranging for its transference to industry. Again it may mean that our industries have been organized on too small a scale to afford the expense of research laboratories. If this last reason be true, then we must conclude that the individual existence in their present form of these small concerns is a serious drag on industrial progress. Is it possible for small-scale undertakings to envisage the larger issues that are vital to the community as a whole? Can they afford to take a long view? The answers to these questions should be sought by a close scrutiny of the present organization of industry and the ends which it is presumed to serve.

Fifty years ago Britain was more advanced technically than any other country. We were the most highly equipped industrial nation in the world. Now we find ourselves sliding down the industrial scale. Why? British workmanship is not discredited;

the craftsmanship of our mechanics remains, with justice, pre-eminent throughout the world. Nor is our science at fault. I should not be seriously challenged if I said that our scientific men are in the forefront in ability. It is freely acknowledged that British scientists have been responsible for the discovery of many fundamental principles which have revolutionized industry the world over. There are many historical reasons why Britain should be losing her industrial pre-eminence, but apart from these, somewhere in between the scientific man and the mechanic there is a gap which tends to render futile scientific ability and mechanical skill. It is the existence of that gap which I have endeavoured to reveal.

6. EVERYBODY A SCIENTIST

"Will an accurate and objective statement of scientific knowledge be sufficient to ensure its general acceptance?" This is one of the most important questions with which educationalists are faced, and not only educationalists, parents, and propagandists, but everyone of us. The answer given to it will decide whether science will be a serious factor in cultural progress. Some clue to the answer can be found in the following illustration.

Suppose I have in my hand a most beautiful rosy apple, and I offer it to you to eat. You ask where I got it. From one of those apple-trees in the graveyard—fruit-trees thrive very well in graveyards, you know. You do not want it? You will not eat it? What is the matter with it? It is as good as any other apple. It has the same acids, the same sugars, the same rosy skin and luscious pulp, the same chemical composition, and yet you will not eat it. Well, I will not press you. Drink this glass of water instead. Where did I get the water from? I can assure you it is very special water indeed—highly purified. I went to-day to a sewage works, and there I saw it being purified by a new biological process. You would hardly believe it to look at it, but it came from a repulsive, evil-smelling mush, and now it is guaranteed absolutely pure, free from bacteria or any form of contamination. What! You will not drink it? Well, you are foolish, it is really cleaner and purer than the stuff that comes out of your tap. If you knew that the water that comes out of your tap, the water that you drink with such relish. . . . What, you do not want to hear? How extraordinary! When I assure you that this water is chemically and biologically pure, you will not drink it, and when I propose to tell you about the water you do drink, you would rather go on drinking than know the truth. It all seems very illogical!

Now, in spite of the ideally scientific attitude I have assumed in this imaginary dialogue, I freely admit that I might find myself unable also to drink that water. I am doubtful, too, about the apple. Please recognize the importance of the admission I have made. You will see that although I am aware of the scientific facts, my actions in the last resort do not seem to be controlled by those facts alone. Indeed, if I had been ignorant of the facts, I might have enjoyed both the apple and the water, but my knowledge called up a feeling of revulsion—they strongly affected my private world.

Now I have been trying to stress the fact that scientific explanation is kept outside this private world, borrowing nothing from it, building its framework independently of individual likes and dislikes. In so far as scientists are people with social ideals, individuals who would like to see the scientific spirit reflected in human development, they proceed on the assumption that if only the facts are set out clearly enough people must necessarily accept them. More than that, they assume that people will not merely agree that the facts are true by saying so in words, but will *behave* as if they believed their truth. Is this a mistaken belief? Do people, in fact, act in accordance with their beliefs as expressed in words? Every scientist and every teacher who would like to see science pull its weight in the community must face up to this problem of getting *conduct* to reflect *belief*. Otherwise belief becomes merely a verbal futility.

I remember one bright winter morning, when I was a student, sitting reading in my room with the sunlight streaming through the window. Presently my landlady entered and quietly pulled the blind, explaining that unless she did so the sun would put out the fire. I was young and argumentative, and immediately tried to find out on what evidence she based her belief. She had, of course, nothing very definite to go on, beyond a very firm belief and a wide experience; but her experience consisted, I discovered, in recognizing that when the blind was drawn the fire always seemed to brighten up. I pointed out that the same would apply to a candle. When the sun shone on it, it seemed to become dimmer, and yet it did not appear to burn any less steadily. The light of the candle was merely obscured, but, like the fire, it became bright again when the blind was drawn. If the fire was being put out by the sun, did it mean that the coal was burning less steadily? There was, of course, no information on the point, and she admitted that until there was such information one could hardly assert that the fire was being put out. I pulled the blind up and showed her that all that apparently happened was that the flames became less distinct, and at the same time the blackness of the coals and the white flakiness of the ash stood out more clearly in contrast. I tried to disabuse her mind of what I conceived, perhaps falsely, to be merely a piece of faulty observation, a prolonged superstition. I thought she appeared convinced. To strengthen the argument I drew from her the admission that she had never discovered that sunlight made a gas fire burn any slower. I settled down to my book with that satis-factory feeling peculiar to the successful propagandist. An hour later I suddenly became aware that she had quietly entered the room and again pulled down the blind! Possibly she suspected that there was a great deal more in the question than I had raised. Possibly there may be. Whatever it was, she evidently had this prior belief, this bias, which induced her to disbelieve the evidence I had set before her eyes, and to continue following a well-established mode of conduct. In fact, she was not prepared to believe the evidence.

Let me remind you of rather an amusing instance of this same disbelief. Alexander Ross in the middle of the seventeenth century attacked Sir Thomas Browne for casting doubt upon the statement, given on the authority of Aristotle, that mice arise from putrefaction. Ross had a firm Aristotelian bias. To question Aristotle's statement as Thomas Browne did, he asserted, was "to question Reason, Sense, and Experience," and he urged doubters to "go to Egypt and see for themselves fields swarming with mice begot of the mud of the Nile." A battery of argument is let loose to support the old tradition. We all hate parting with our superstitions.

There is another illustration in the comet of 1712. Whiston, the mathematical divine, had predicted on good evidence that it would become visible on Wednesday, October 14th, at 5.5 a.m., and that in accordance with an old superstition the world would be destroyed by fire on the following Friday morning. The comet duly arrived to time, and panic spread in the belief that the rest of the prophecy would also be correct. People crowded on to barges and boats on the Thames, feeling that on the water they would be safest. Over one hundred clergymen, it is said, went over to Lambeth Palace to request that special prayers should be said for the emergency. Shares fell heavily. People rushed to take their savings out of the bank. Whiston, of course, had no evidence for this part of his prediction. Scientist enough to appreciate the meaning of an astronomical calculation, he yet held tenaciously to a piece of superstition. That, of course, was in the seventeenth and eighteenth centuries. Are we more advanced now? How many of our superstitious beliefs have we really discarded?

The accumulation of scientific knowledge alone is no guarantee of the purity of our beliefs. Is science itself free from superstitions? One is inclined to doubt it if one looks at the great scientific figures of the past and inquires into the nature of the beliefs they professed. Copernicus, who disestablished the Earth from its splendid position at the centre of the universe, believed that the planets were urged on in their courses by propelling angels. Kepler drew horoscopes. Newton applied his mathematical genius to working out the astrological predictions in the Book of Daniel. Boyle, one of the founders of the Royal Society, believed in the doctrine of the transmutation of metals into gold, so did Newton and Leibniz, co-inventors of the differential calculus. Priestley, who discovered oxygen, could scarcely understand its significance, so intensely did he believe in that mystical fluid Phlogiston which then obsessed scientists.

Even in more modern times scientists have become so accustomed to a theory that they tend to develop a superstitious belief in its truth. The belief in the ether is a case in point. Again, we have all met the intelligent Westerner who, having lived in India for several years, assures us that the famous disappearing rope trick actually takes place. He says that the Indian boy really disappears. Nor need one go to the Orient for tales of such credulity. How many people avoid walking under a ladder, cast salt over their left shoulder, and jokingly touch wood at the suggestion of danger? People still have a sneaking half-belief in mascots and amulets, charms and talismans, and hate to part with them lest bad luck befall. How many people are wearing at this moment some little luck-bringer with which they would not part. And when bad luck comes do they discard their little black pig, or swastika, or elephant's hair ring, or shark's tooth? How many refuse to wear pearls on a Friday? How many believe in the "Laws of Juridical Astrology?" How many could have their fortunes told and bear the indications of an evil fate with complete equanimity? Is not the swagger of disbelief an indication of a little nervousness?

We must not fall into the error, therefore, of presuming that man, in his present stage of evolution, can be satisfied by science alone or by cold reason. We are complex creatures wrapped up in our past, and we must not expect our conduct to be dictated by intellectual ideas alone. Common sense cries for consistent and reasonable actions, but how many of us are consistent and reasonable?

An important communal issue is raised. If we are each to continue to act according to our particular prejudices, we are going to find it hard to live together in peace. Since we *must* live together we shall have to find a common basis of conduct arising out of accepted and agreed beliefs. Belief resting on scientific fact is, so far, the sole method of approach to knowledge which enables unanimity of assent to be won.

It is by education that we usually try to teach the difficult task of living together. If education is to fulfil its task it will require to be permeated with the scientific spirit, it will require to eliminate from its teachings all superstitions and all beliefs that cannot stand critical examination. Every question raised by the child will have to be taken and made an excuse for examining the evidence for the answer given. Throughout its career the child will have to be accustomed to sifting evidence. It must be taught to appreciate the meaning of knowledge and the perspective of man's forward groping. Education must be built upon a scientific basis. I do not suggest that teachers are not at present carrying out a difficult task in an admirable manner, but their endeavours are often limited by the superstitions of the child's environment. I am not now discussing technical education. My plea is for something fundamentally cultural. As civilization becomes more mechanized and scientific, and as communal life comes to rely more closely on the steady functioning of scientific processes, it is essential that educational methods should develop on parallel lines. Just as old machinery becomes obsolete and is scrapped, so false traditions and baseless superstitions must be discarded and eliminated from the social heritage.

It is largely a question of early habits. Unless we early instil critical habits into our children, unless we encourage them to call for evidence on all conceivable occasions, they will rapidly adopt a habit of simply *believing*, duly followed by "believing behaviour." Thus superstitions persist and remain in the common stock of traditional action. It is obvious that if ill-founded beliefs are to be eliminated verbal exposure alone is not enough. The behaviour called forth by the superstition must also receive attention. It may not be difficult to demonstrate that belief in the unluckiness of the number 13 has no foundation, but how are you going to prevent people from *acting* upon a belief that 13 is unlucky? How are you going to make them sit down 13 at a table?

Educationalists are faced with a twofold problem. In the first place teachers must see to it that throughout the whole course of the child's education a critical attitude is adopted to the question of *evidence*. In the second place teachers must set out to *condition* behaviour. It is not sufficient to explode the superstition; the explosion has also to shake the child of a *habit*. And that is not easy, for by the time children arrive at school many superstitions and prejudices, taboos and traditions, are already well established. A *bias* has been introduced into the child's outlook from its tenderest years. How to adjust educational practice to meet these difficulties is a matter for serious study, and teachers themselves will require to shed many superstitions in the investigation.

In a thousand ways Science has already been called in to the help of Education—books, pictures, films, gramophones, wireless, are all aids to learning. But these things are the machinery of science; the spirit of science has yet to be liberated for educational service and instilled into social relations. It is a problem that calls for the enterprise and initiative of our generation of teachers and thinkers. By striving to permeate social life with the spirit of critical foresight, by seeking to guide conduct with accurate knowledge, Science may yet carve out a new future for mankind.

PART II

WHAT IS MAN?

JULIAN HUXLEY

and

JOHN R. BAKER

JULIAN HUXLEY

1. MAN AS A RELATIVE BEING

DURING the present century we have heard so much of the revolutionary discoveries of modern physics that we are apt to forget how great has been the change in the outlook due to biology. Yet in some respects this has been the more important. For it is affecting the way we think and act in our everyday existence. Without the discoveries and ideas of Darwin and the other great pioneers in the biological field, from Mendel to Freud, we should all be different from what we are. The discoveries of physics and chemistry have given us an enormous control over lifeless matter and have provided us with a host of new machines and conveniences, and this certainly has reacted on our general attitude. They have also provided us with a new outlook on the universe at large: our ideas about time and space, matter and creation, and our own position in the general scheme of things, are very different from the ideas of our grandfathers.

Biology is beginning to provide us with control over living matter—new drugs, new methods for fighting disease, new kinds of animals and plants. It is helping us also to a new intellectual outlook, in which man is seen not as a finished being, single lord of creation, but as one among millions of the products of an evolution that is still in progress. But it is doing something more. It is actually making us different in our natures and our biological behaviour. I will take but three examples.

The application of the discoveries of medicine and physiology is making us healthier: and a healthy man behaves and thinks differently from one who is not so healthy. Then the discoveries of modern psychology have been altering our mental and emotional life, and our system of education: taken in the mass, the young people now growing up feel differently, and will therefore act differently, about such vital matters as sex and marriage, about jealousy, about freedom of expression, about the relation between parents and children. And as a third example, as a race we are changing our reproductive habits: the idea and the practice of deliberate birth-control has led to fewer children. People living in a country of small families and a stationary or decreasing population will in many respects *be* different from people in a country of large families and an increasing population.

This change has not been due to any very radical new discoveries made during the present century. It has been due chiefly to discoveries which were first made in the previous century, and are at last beginning to exert a wide effect. These older discoveries fall under two chief heads. One is Evolution —the discovery that all living things, including ourselves, are the product of a slow process of development which has been brought about by natural forces, just as surely as has to-day's weather or last month's high tides. The other is the sum of an enormous number of separate discoveries which we may call physiological, and which boil down to this: that all living things, again including ourselves, work according to regular laws, in just the same way as do non-living things, except that living things are much more complicated. The old idea of "vital force" has been driven back and back until there is hardly any process of life where it can still find any foothold. Looked at objectively and scientifically, a man is an exceedingly complex piece of chemical machinery. This does not mean that he cannot quite legitimately be looked at from other points of view—subjectively, for instance; what it means is that so far as it goes, this scientific point of view is true, and not the point of view which ascribed human activities to the working of a vital force quite different from the forces at work in matter which was not alive.

Imagine a group of scientists from another planet, creatures with quite a different nature from ours, who had been dispassionately studying the curious objects called human beings for a number of years. They would not be concerned about what we men felt we were or what we would like to be, but only about getting an objective view of what we actually were and why we were what we were. It is that sort of picture which I want to draw for you.

Our Martian scientists would have to consider us from three main viewpoints if they were to understand much about us. First they would have to understand our physical construction, and what meaning it had in relation to the world around and the work we have to do in it. Secondly, they would have to pay attention to our development and our history. And thirdly, they would have to study the construction and working of our minds. Any one of these three aspects by itself would give a very incomplete picture of us.

An ordinary human being is a lump of matter weighing between 50 and 100 kilograms. This living matter is the same matter of which the rest of the earth, the sun, and even the most distant stars and nebulae are made. Some elements which make up a large proportion of living matter, like hydrogen and especially carbon, are rare in the not-living parts of the earth; and others which are abundant in the earth are, like iron, present only in traces in living creatures, or altogether absent, like aluminium or silicon. None the less, it is the same matter. The chief difference between living and non-living matter is the complication of living matter. Its elements are built up into molecules much bigger and more elaborate than any others known, often containing more than a thousand atoms each. And, of course, living matter has the property of self-reproduction; when supplied with the right materials and in the right conditions, it can build up matter which is not living into its own complicated patterns.

Life, in fact, from the "public" standpoint, which Professor Levy has stressed as being the only possible standpoint for science, is simply the name for the various distinctive properties of a particular group of very complex chemical compounds. The most important of these properties are, first, feeding, assimilation, growth, and reproduction, which are all aspects of the one quality of self-reproduction; next, the capacity for reacting to a number of kinds of changes in the world outside—to stimuli, such as light, heat, pressure, and chemical change; then the capacity for liberating energy in response to these stimuli, so as to react back again upon the outer world—whether by moving about, by constructing things, by discharging chemical products, or by generating light or heat; and finally the property of variation. Self-reproduction is not always precisely accurate, and the new substance is a little different from the parent substance which produced it.

The existence of self-reproduction on the one hand and variation on the other automatically leads to what Darwin called Natural Selection. This is a sifting process, by which the different new variations are tested out against the conditions of their existence, and in which some succeed better than others in surviving and in leaving descendants. This blind process slowly but inevitably causes living matter to change—in other words, it leads to evolution.

There may be other agencies at work in guiding the course of evolution; but it seems certain that Natural Selection is the most important.

The results it produces are roughly as follows. It *adapts* any particular stream of living matter more or less completely to the conditions in which it lives. As there are innumerable different sets of conditions to which life can be adapted, this has led to an increasing diversity of life, a splitting of living matter into an increasing number of separate streams. The final tiny streams we call species; there are perhaps a million of them now in existence. This adaptation is progressive; any one stream of life is forced to grow gradually better and better adapted to some particular condition of life. We can often see this in the fossil records of past life. For instance, the early ancestors of lions and horses about 50 million years ago were not very unlike, but with the passage of time one line grew better adapted to catching and eating large prey, the other grew better adapted to grass-eating and running away from enemies. And finally natural selection leads to general progress; there is a gradual raising of the highest level attained by life. The most advanced animals are those which have changed their way of life and adapted themselves to new conditions, thus taking advantage of biological territory hitherto unoccupied. The most obvious example of this was the invasion of the land. Originally all living things were confined to life in water, and it was not for hundreds of millions of years after the first origin of life that plants and animals managed to colonize dry land.

But progress can also consist in taking better advantage of existing conditions: for instance, the mammal's biological inventions, of warm blood and of nourishing the unborn young within the mother's body, put them at an advantage over other inhabitants of the land; and the increase in size of brain which is man's chief characteristic has enabled him to control and exploit his environment in a new and more effective way, from which his pre-human ancestors were debarred.

It follows from this that all animals and plants that are at all highly developed have a long and chequered history behind them, and that their present can often not be properly understood without an understanding of their past. For instance, the tiny hairs all over our own bodies are a reminder of the fact that we are descended from furry creatures, and have no significance except as a survival.

Let us now try to get some picture of man in the light of these ideas. The continuous stream of life that we call the human race is broken up into separate bits which we call individuals. This is true of all higher animals, but is not necessary: it is a convenience. Living matter has to deal with two sets of activities: one concerns its immediate relations with the world outside it, the other concerns its future perpetuation. What we call an individual is an arrangement permitting a stream of living matter to deal more effectively with its environment. After a time it is discarded and dies. But within itself it contains a reserve of potentially immortal substance, which it can hand on to future generations, to produce new individuals like itself. Thus from one aspect the individual is only the casket of the continuing race; but from another the achievements of the race depend on the construction of its separate individuals.

The human individual is large as animal individuals go. Size is an advantage if life is not to be at the mercy of small changes in the outer world: for instance, a man the size of a beetle could not manage to keep his temperature constant. Size also goes with long life: and a man who only lived as long as a fly could not learn much. But there is a limit to size; a land animal much bigger than an elephant is not, mechanically speaking, a practical proposition. Man is in that range of size, from 100 lb. to a ton, which seems to give the best combination of strength and mobility. It may be surprising to realize that man's size and mechanical construction are related to the size of the earth which he inhabits; but so it is. The force of gravity on Jupiter is so much greater than on our own planet, that if we lived there our skeletons would have to be much stronger to support the much increased weight which we would then possess, and animals in general would be more stocky; and conversely, if the earth were only the size of the moon, we could manage with far less expenditure of material in the form of bone and sinew, and should be spindly creatures.

Our general construction is determined by the fact that we are made of living matter, must accordingly be constantly passing a stream of fresh matter and energy through ourselves if we are to live, and must as constantly be guarding ourselves against danger if we are not to die. About 5 per cent of ourselves consists of a tube with attached chemical factories, for taking in raw materials in the shape of food, and converting it into the form in which it can be absorbed into our real interior. About 2 per cent consists in arrangements—windpipe and lungs —for getting oxygen into our system in order to burn the food materials and liberate energy. About 10 per cent consists of an arrangement for distributing materials all over the body—the blood and lymph, the tubes which hold them and the pump which drives them. Much less than 5 per cent is devoted to dealing with waste materials produced when living substance breaks down in the process of producing energy to keep our machinery going— the kidneys and bladder and, in part, the lungs and skin. Over 40 per cent is machinery for moving us about—our muscles; and nearly 20 per cent is needed to support us and to give the mechanical leverage for our movements—our skeleton and sinews. A relatively tiny fraction is set apart for giving us information about the outer world—our sense organs. And there is about 3 per cent to deal with the difficult business of adjusting our behaviour to what is happening around us. This is the task of the ductless glands, the nerves, the spinal cord and the brain; our conscious feeling and thinking is done by a small part of the brain. Less than 1 per cent of our bodies is set aside for reproducing the race. The remainder of our body is concerned with special functions like protection, carried out by the skin (which is about 7 per cent of our bulk) and some of the white blood corpuscles; or temperature regulation, carried out by the sweat glands. And nearly 10 per cent of a normal man consists of reserve food stores in the shape of fat.

Other streams of living matter have developed quite other arrangements in relation to their special environment. Some have parts of themselves set aside for manufacturing electricity, like the electric eel; or light, like the firefly. Some, like certain termites, are adapted to live exclusively on wood; others, like lions, exclusively on flesh; others, like cows, exclusively on vegetables. Some, like boa-constrictors, only need to eat every few months; others, like parasitic worms, need only breathe a few hours a day; others, like some desert gazelles, need no water to drink. Many cave animals have no eyes; tapeworms have no mouths or stomachs; and so on and so forth. And all these peculiarities, including those of our own construction, are related to the kind of surroundings in which the animal lives.

This relativity of our nature is perhaps most clearly seen in regard to our senses. The ordinary man is accustomed to think of the information given by his senses as absolute. So it is—for him; but not in the view of our Martian scientist. To start with, the particular senses we possess are not shared by many other creatures. Outside backboned animals, for instance, very few creatures can hear at all; a few insects and perhaps a few crustacea probably exhaust the list. Even fewer animals can see colours; apparently the world as seen even by most mammals is a black and white world, not a coloured world. And the majority of animals do not even see at all in the sense of being given a detailed picture of the world around. Either they merely distinguish light from darkness, or at best can get a blurred image of big moving objects. On the other hand, we are much worse off than many other creatures—dogs, for instance, or some moths—in regard to smell. Our sense of smell is to a dog's what an eye capable of just distinguishing big moving objects is to our own eye.

But from another aspect, the relativity of our senses is even more fundamental. Our senses serve to give us information about changes outside our bodies. Well, what kind of changes are going on in the outside world? There are ordinary mechanical changes: matter can press against us, whether in the form of a gentle breeze or a blow from a poker. There are the special mechanical changes due to vibrations passing through the air or water around us—these are what we hear. There are changes in temperature—hot and cold. There are chemical changes—the kind of matter with which we are in contact alters, as when the air contains poison gas, or our mouth contains lemonade. There are electrical changes, as when a current is sent through a wire we happen to be touching.

And there are all the changes depending on what used to be called vibrations in the ether. The most familiar of these are light-waves; but they range from the extremely short waves that give cosmic rays and X-rays, down through ultra-violet to visible light, on to waves of radiant heat, and so on to the very long Hertzian waves which are used in wireless. All these are the same kind of thing, but differ in wave-length.

Now of all these happenings, we are only aware of what appears to be a very arbitrary selection. Mechanical changes we are aware of through our sense of touch. Air-vibrations we hear; but not all of them—the small wave-lengths are pitched too high for our ears, though some of them can be heard by other creatures, such as dogs and bats. We have a heat sense and a cold sense, and two kinds of chemical senses for different sorts of chemical changes—taste and smell. But we possess no special electrical sense—we have no way of telling whether a live rail is carrying a current or not unless we actually touch it, and then what we feel is merely pain.

The oddest facts, however, concern light and kindred vibrations. We have no sense organs for perceiving X-rays, although they may be pouring into us and doing grave damage. We do not perceive ultra-violet light, though some insects, like bees, can see it. And we have no sense organs for Hertzian waves, though we make machines—wireless receivers —to catch them. Out of all this immense range of vibrations, the only ones of which we are aware through our senses are radiant heat and light. The waves of radiant heat we perceive through the effect which they have on our temperature sense organs; and the light-waves we see. But what we see is only a single octave of the light-waves, as opposed to ten or eleven octaves of sound-waves which we can hear.

This curious state of affairs begins to be comprehensible when we remember that our sense organs have been evolved in relation to the world in which our ancestors lived. In nature, for instance, large-scale electrical changes hardly occur. The only exceptions are electrical discharges such as lightning, and they act so capriciously and violently that to be able to detect them would be no advantage. The same is true of X-rays. The amount of them knocking about under normal conditions is so small that there is no point in having sense organs to tell us about them. Wireless waves, on the other hand, are of such huge wave-lengths that they go right through living matter without affecting it. Even if they were present in nature, there would be no obvious way of developing a sense organ to perceive them.

As regards light, there seem to be two reasons why our eyes are limited to seeing only a single octave of the waves. One is that of the ether vibrations raying upon the earth's surface from the sun and outer space, the greatest amount is centred in this region of the spectrum; the intensity of light of higher or lower wave-lengths is much less, and would only suffice to give us a dim sensation. Our greatest capacity for seeing is closely adjusted to the amount of light to be seen. The other is more subtle, and has to do with the properties of light of different wave-lengths. Ultra-violet light is of so short a wavelength that much of it gets scattered as it passes through the air, instead of progressing forward in straight lines. Hence a photograph which uses only the ultra-violet rays is blurred and shows no details of the distance. A photograph taken by infra-red light, on the other hand, while it shows the distant landscape very well, over-emphasizes the contrast between light and shade in the foreground. Leaves and grass reflect all the infra-red, and so look white, while the shadows are inky-black, with no gradations. The result looks like a snowscape. An eye which could only see the ultra-violet octave would see the world as in a fog; and one which could see only the infra-red octave would find it impossible to pick out lurking enemies in the jet-black shadows.

The particular range of light to which our eyes are attuned gives the best-graded contrast.

Then of course there is the pleasant or unpleasant quality of a sensation; and this, too, is in general related to our way of life. I will take one example. Both lead acetate and sugar taste sweet; the former is a poison, but very rare in nature; the latter is a useful food, and common in nature. Accordingly we most of us find a sweet taste pleasant. But if lead acetate were as common in nature as sugar, and sugar as rare as lead acetate, it is safe to prophesy that we should find sweetness a most horrible taste, because we should only survive if we spat out anything which tasted sweet.

Now let us turn to another feature of man's life which would probably seem exceedingly queer to a scientist from another planet—sex. We are so used to the fact that our race is divided up into two quite different kinds of individuals, male and female, and that our existence largely circles round this fact, that we rarely pause to think about it. But there is no inherent reason why this should be so. Some kinds of animals consists only of females; some, like ants, have neuters in addition to the two sexes; some plants are altogether sexless.

As a matter of fact, the state of affairs as regards human sex is due to a long and curious sequence of causes. The fundamental fact of sex has nothing to do with reproduction; it is the union of two living cells into one. The actual origin of this remains mysterious. Once it had originated, however, it proved of biological value, by conferring greater variability on the race, and so greater elasticity in meeting changed conditions. That is why sex is so nearly universal. Later, it was a matter of biological convenience that reproduction in higher animals became indissolubly tied up with sex. Once this had happened, the force of natural selection in all its intensity became focused on the sex instinct, because in the long run those strains which reproduce themselves abundantly will live on, while those which do not do so will gradually be supplanted.

A wholly different biological invention, the retention of the young within the mother's body for protection, led to the two sexes becoming much more different in construction and instincts than would otherwise have been the case. The instinctive choice of a more pleasing as against a less pleasing mate—what Darwin called sexual selection—led to the evolution of all kinds of beautiful or striking qualities which in a sexless race would never have developed. The most obvious of such characters are seen in the gorgeous plumage of many birds; but sexual selection has undoubtedly modelled us human beings in many details—the curves of our bodies, the colours of lips, eyes, cheeks, the hair of our heads, and the quality of our voices.

Then we should not forget that almost all other mammals and all birds are, even when adult, fully sexed only for a part of the year; after the breeding season they relapse into a more or less neuter state. How radically different human life would be if we too behaved thus! But man has continued an evolutionary trend begun for some unknown reason among the monkeys, and remains continuously sexed all the year round. Hunger and love are the two primal urges of man: but by what a strange series of biological steps has love attained its position!

We could go on enumerating facts about the relativity of man's physical construction; but time is short, and I must say a word about his mind. For that too has developed in relation to the conditions of our life, present and past. Many philosophers and theologians have been astonished at the strength of the feeling which prompts most men and women to cling to life, to feel that life is worth living, even in the most wretched circumstances. But to the biologist there is nothing surprising in this. Those men (and animals) who have the urge to go on living strongly developed will automatically survive and breed in greater numbers than those in whom it is weak. Nature's pessimists automatically eliminate themselves, and their pessimistic tendencies, from the race. A race without a strong will to live could no more hold its own than one without a strong sexual urge.

Then again man's highest impulses would not exist if it were not for two simple biological facts— that his offspring are born helpless and must be protected and tended for years if they are to grow up, and that he is a gregarious animal. These facts make it biologically necessary for him to have well-developed altruistic instincts, which may and often do come into conflict with his egoistic instincts, but are in point of fact responsible for half of his attitude towards life. Neither a solitary creature like a cat or a hawk, nor a creature with no biological responsibility towards its young, like a lizard or a fish, could possibly have developed such strong altruistic instincts as are found in man.

Other instincts appear to be equally relative. Everyone who has any acquaintance with wild birds and animals knows how much different species differ in temperament. Most kinds of mice are endowed with a great deal of fear and very little ferocity; while the reverse is true of various carnivores like tigers or Tasmanian devils. It would appear that the amounts of fear and anger in man's emotional make-up are greater than his needs as a civilized being, and are survivals from an earlier period of his racial history. In the dawn of man's evolution from apes, a liberal dose of fear was undoubtedly needed if he was to be preserved from foolhardiness in a world peopled by wild beasts and hostile tribes, and an equally liberal dose of anger, the emotion underlying pugnacity, if he was to triumph over danger when it came. But now they are on the whole a source of weakness and maladjustment.

It is often said that you cannot change human

nature. But that is only true in the short-range view. In the long run, human nature could as readily be changed as feline nature has actually been changed in the domestic cat, where man's selection has produced an amiable animal out of a fierce ancestral spit-fire of a creature. If, for instance, civilization should develop in such a way that mild and placid people tended to have larger families than those of high-strung or violent temperament, in a few centuries human nature would alter in the direction of mildness.

But it is not only in regard to instincts and feelings that our mind bears the stamp of the world around. Bergson, the French philosopher, has gone as far as to suggest that the very way our thinking processes work is adapted to practical needs. To satisfy the primary needs of life, man must handle and deal with definite, separate material objects; to get a general picture of the continuity of things in space or time is not so pressing. In general it is what we call intellect which thinks in the first way, about separate objects; and what we call intuition which thinks in the second way, about whole situations. Bergson points out that the evolution of our minds has been largely determined by the practical necessity for thinking in the first way, and that the way men think is not the only way in which thinking could be done. On the contrary, in a different kind of world, organisms might develop in which most thinking was intuitive.

If these ideas of Bergson's are perhaps a little

speculative, they are none the less worth reflecting on, as showing how the human mind is doubly imprisoned—it is imprisoned in its own way of thinking and feeling, and this way of thinking and feeling is itself in a sense imprisoned in the material world about it. When we come to another fundamental property of our minds, however, we are on firmer ground. I mean our capacity for forgetting. This is usually taken to be a natural property or at least a natural imperfection of mind. And a certain amount of our forgetting does seem to be due to this. A great deal, however, quite definitely does not, but owes its existence to the practical needs of our life.

To a large extent we forget what it is convenient for our own purpose to forget. If we ever do get a chance to see ourselves as others see us, it is generally a shock to find how many inconvenient facts about ourselves and our actions, which are all too prominent in the minds of others, have been forgotten by ourselves.

Pavlov has shown how even dogs can be made to have nervous breakdowns by artificially generating in their minds conflicting urges to two virtually exclusive kinds of action; and we all know that the same thing, on a higher level of complexity, happens in human beings. But a nervous breakdown puts an organism out of action for the practical affairs of life, quite as effectively as does an ordinary infectious disease. And just as against physical germ-diseases we have evolved a protection in the shape of the immunity reactions of our blood, so we have evolved oblivion as protection against the mental diseases arising out of conflict. For, generally speaking, what happens is that we forget one of the two conflicting ideas or motives. We do this either by giving the inconvenient idea an extra kick into the limbo of the forgotten, which psychologists call suppression, or else, when it refuses to go so simply, by forcibly keeping it under in the sub-conscious, which is styled repression. For details about suppression and repression and their often curious and sometimes disastrous results I must ask you to refer to any modern book on psychology. All I want to point out here is that a special mental machinery has been evolved for putting inconvenient ideas out of consciousness, and that the contents and construction of our minds are different in consequence.

Our current ideas, our feelings, our scientific discoveries, our laws, even our religions are moulded by the social environment of the period. We live in a more or less scientific age : it is all but impossible for us to know what it would feel like to live in a community which believed chiefly in magic. It is equally impossible for us, living in an age of nationalism, to look forward and know how people would feel and behave in a unified, super-nationalist world. We find it impossible to understand how our great-grandfathers tolerated child-labour and slavery; it is likely that our great-grandchildren will find it

I

equally impossible to understand how we tolerated capital punishment or our present penal system.

But I have said enough, I hope, to give you some idea of what is implied by calling man a relative being. It implies that he has no real meaning apart from the world which he inhabits. Perhaps this is not quite accurate. The mere fact that man, a portion of the general stuff of which the universe is made, can think and feel, aspire and plan, is itself full of meaning, but the precise way in which man is made, his physical construction, the kinds of feelings he has, the way he thinks, the things he thinks about, everything which gives his existence form and precision—all this can only be properly understood in relation to his environment. For he and his environment make one interlocking whole.

The great advances in scientific understanding and practical control often begin when people begin asking questions about things which up till then they have merely taken for granted. If humanity is to be brought under its own conscious control, it must cease taking itself for granted, and, even though the process may often be humiliating, begin to examine itself in a completely detached and scientific spirit.

JOHN R. BAKER

2. OUR PLACE IN NATURE

PROFESSOR JULIAN HUXLEY has reminded you that as a result of the discoveries made in the last thirty years, we do not regard man exactly as we did at the end of the nineteenth century, and I now propose to give some account of the state of knowledge on this subject. Naturally there will be accounts of observations and experiments which not every one will be in a position to repeat, but if you doubt some comparison, for instance, between a gorilla's skull and a man's, you should go to a museum and look for yourself. I hope that you will look in a zoo or museum at the apes to which I shall refer. Familiarize yourself with the gorilla, chimpanzee, orang-utan and gibbons, so that you could not mistake one for another. Gorillas are rare in zoos in Great Britain, as they do not live well in captivity, but you can see them in museums, and there are splendid specimens in the Natural History Museum in London. You go up the centre stairs, turn to the right past the giraffes, up the next flight of stairs, and then you will find them through a door on the right. At the London Zoo it is very easy to find the apes, as their house is the very first one you come to when you go in by the main entrance. The chimpanzee and orang-utans are on the right as you enter and the gibbons on the right at the far end. If you do not live in London, there is sure to be a museum where you can see the chimpanzee, and there are, of course, other excellent zoos in various parts of the country. What I have to say will mean much more to you if you have a concrete idea of what the animals are like, even if you cannot make an elaborate study of their anatomy.

Where is man placed in the animal kingdom to-day? He is obviously a *Mammal*, that is, he stands in the same group as rabbits and mice and cats and dogs and horses and cattle, which all have hair, are born in a fairly advanced condition, instead of being laid as eggs, and which are all suckled by their mothers after birth. There are, of course, many more characters which we have in common with the other Mammals. Now the Mammals are divided into various *Orders*; for instance, the Rodents (the gnawing animals, rabbits and mice and guinea-pigs), the Carnivores (flesh-eating animals, like cats and dogs and bears), the Ungulates (the hoofed animals, like horses and rhinoceroses and cattle and sheep and camels and giraffes). Man quite definitely belongs in the same order as the monkeys, the order *Primates*, or nailed Mammals. These are Mammals with nails on their fingers instead of hoofs or claws, with two teeth on each side of each jaw before the canines, with the eye surrounded by a ring of bone, with well-developed collar-bones, nearly always with ten fingers and ten toes, the thumb being opposable to the other fingers, and with two milk-glands on the chest.

Now where does he stand among the Primates? First of all we can separate off the Lemurs, and say definitely that he does not belong there. The Lemurs are Primates, but very primitive ones, with rather foxy faces, quite unlike monkeys in external appearance. Also they differ from monkeys and ourselves in not having the eye-cavity lined with bone all round behind, and also they have a claw instead of a nail on their second toe. Look out for that claw in the little Lemur house at the London Zoo, but it is not always easy to see, as the animals often sit firmly on their hind feet as though they did not want you to see.

Not counting the Lemurs, we have five families of Primates. These are the Marmosets, South American Monkeys, Old World Monkeys, the Apes, and Men. By the Apes I mean the Gibbons, Orang-utan, Chimpanzee, and Gorilla. Some of the monkeys are often called Apes, but I think it is best to restrict the term in this way. Now to which of the other families is man most closely allied? Not to the Marmosets, obviously, for they are curious little Primates which cannot oppose their thumbs to their fingers and have claws on most of their toes and fingers, instead of nails. The South American monkeys—that is to say, the monkeys with pre-hensile tails—are very different from man. The usual organ-grinder's monkey is one of these. Look at their widely separated nostrils. If you can obtain a skull, count the teeth. You will find they have six grinding teeth on each side of each jaw.

The Old World monkeys, with non-prehensile tails, are much more like man, for their nostrils are close together and they have five grinding teeth on each side of each jaw, just as we have. Nevertheless, they have certain striking differences. Their grinding teeth are somewhat elongated from front to back, instead of being squarish, as ours are. Also many of them have curious swellings on their buttocks, and many have pouches in their cheeks in which they store food. Then, again, their breast-bone is narrow, and they have no appendix and they usually have tails.

Now I have mentioned a number of ways in which the Old World monkeys differ from man, *and in every one of these points the Apes resemble man.* There can be no doubt from comparative anatomy that the apes are closer to man than any other animals. We may not like to come next to the gorilla, chimpanzee, orang-utan and gibbon, but in our anatomy we undoubtedly do. I must repeat that the Apes resemble us in having squarish grinding teeth, a broad breast-bone, an appendix, no swellings on the buttocks, no cheek-pouches and no tail. Further, they often walk on their hind legs. The gibbon walks absolutely erect.

We do not come in the same family as the Apes, because we do differ from them in certain important respects. First and foremost we have our big toe, which, as its name indicates, is the largest of our toes, which it never is in the apes. Then our big toe cannot be opposed to our other toes, and our legs are longer than our arms, and our jaws stick forward less, while our chin sticks forward instead of receding, and our canine teeth do not project beyond the others, and we have not got great bony ridges above our eyes, and we have far less hair, and last, but by no means least, our brain is very much bigger. It is very nearly the biggest in the whole animal kingdom, although we are so small compared with many animals. It is far larger than that of a cow or horse, and more than twice as big as the largest ape's brain. It is only exceeded in size by the brain of the elephant and certain whales.

These differences suffice to place us in a separate family, but probably we are more closely allied to the apes than the apes are to the Old World monkeys.

I want to give you some idea of what it means to say that we are closely allied to the apes, but in a separate family. It means that in our anatomy we are about as different from apes as antelopes are from deer, or as hyenas are from dogs. You must not suppose that that sort of comparison is very exact, but it gives an idea of the scale of the differences.

Now we have investigated the anatomy of man and his relations a little, but we have not considered the working of their bodies—that is, their physiology.

Does man's body work in much the same way as that of apes, or are there radical differences, which show that man is not so closely allied to them as their anatomy would make us think?

A lot of work has been done on this subject recently, but it is incomplete, simply because it is so difficult and expensive to use apes as laboratory animals.

You know, of course, that gout is caused by uric acid, which has an unpleasant way of collecting in joints. Now what is this uric acid? It is a product of the nuclei of the cells in the food which you eat, and of the nuclei of the cells of your own body. All Mammals make uric acid in their bodies, but most of them turn about half of it into another substance, called allantoin, which does not get lodged in joints. The Old World monkeys do this and so they are most unlikely to suffer from gout. But man has no capacity whatever of changing uric acid into anything else, and so it must either be excreted as such or else stored up in the joints in a most painful manner. I do not think that anyone has studied the gorilla or orang-utan in this connection, but Hunter has studied the chimpanzee and he has found that it exactly resembles man and differs from the Old World monkeys and lower Mammals. It has no capacity of changing uric acid into allantoin. This agrees with our conclusion from anatomy that man is more closely allied to the apes than the apes are to monkeys.

Now let us take another branch of physiology and see how man compares with apes. Perhaps you have had occasion to give blood to someone else by blood-transfusion. If you have, you will remember that it is not everyone who has the right sort of blood to give to the person who happens to need it. If you have the wrong sort of blood for a certain patient, then his blood will destroy the blood corpuscles which you give him. Your blood corpuscles will all stick together in clumps and finally degenerate, and they will not be of any use to him. Nevertheless your blood may be perfectly suitable for transfusion into the blood of somebody else, and if you are in urgent need of blood yourself one day, then the saving of your life by blood-transfusion will depend on knowledge gained in experiments like the ones I am going to describe. These actual experiments were performed in America not many years ago by Landsteiner and Miller.

Suppose you take some blood of a macaque monkey—which is an ordinary sort of Old World monkey—and inject it into a rabbit. What happens? The blood of the rabbit gets the property of being able to make macaque blood corpuscles stick together. You can take some of this rabbit's blood, and even if you dilute it enormously, it still possesses this power of making macaque blood corpuscles stick together. It has the same effect on baboon blood. Now macaques and baboons are rather closely related: they are in the same family. So perhaps it is not very surprising that their blood corpuscles behave in the same way when put into this rabbit's blood. Even when enormously diluted, this rabbit's blood causes baboon blood corpuscles to stick together.

What about chimpanzee blood? Suppose you take some of it and mix it with some blood from the same rabbit as before, which was previously injected with macaque blood. Now what happens? Will it cause the chimpanzee's blood corpuscles to stick together? Scarcely at all. It is clear that chimpanzee's blood is very different from that of the macaque and baboon.

What about human blood? It is the same as with the chimpanzee's. The rabbit's blood, which is so fatal to the blood corpuscles of the macaque and baboon, has scarcely any effect.

It is clear that in their blood reactions man and the chimpanzee are equally distantly related to the macaque and baboon.

Take another rabbit and inject human blood into it instead of macaque's. Now this rabbit's blood acquires the property of making *human* blood corpuscles stick together. Let us take some of this rabbit's blood and try mixing chimpanzee's blood with it. The chimpanzee's blood corpuscles stick together. Evidently chimpanzee's blood is very much like that of man.

Is it exactly the same? Can we distinguish it by tests of this sort? Take some of this last rabbit's blood, which causes both human and chimpanzee corpuscles to stick together. Put it in a glass vessel and go on adding chimpanzee corpuscles to it until all the substance which causes chimpanzee corpuscles to stick together is used up. Now filter it, and you have got rabbit's blood which has practically lost its capacity to make chimpanzee's corpuscles stick together. Now make the crucial test. Add human blood. It still causes them to stick together. Even when enormously diluted, it still retains that property. So you see you *can* distinguish chimpanzee's blood from man's by tests of this sort, but only in rather a roundabout way.

There is one way in which man differs very much from most wild animals, and that is in not having a breeding season. It is true, of course, that more births occur at one time of year than another, and this is especially so among the Eskimos, but on the whole we can say that the human race is without a special season. This was thought to be a peculiarity of man, and perhaps a result of what we might call domestication. Several domestic animals have lost their breeding season as a result of domestication. Unlike their wild relations, the cow and the pig breed all the year round. Mr. Zuckermann has been looking into this matter lately, and he has come to the conclusion that man, after all, is not so peculiar in not having a breeding season, because he finds that many Old World monkeys lack one, and breed at any time, producing young ones at all seasons of the year.

I must just mention one theory which Professor Wood Jones has lately been suggesting. He thinks that we are descended from animals like Tarsius, and not from an ape-like ancestor at all. Tarsius is a little lemur which is very different from all the other lemurs. Go and look at him in a museum, for I think he does not exist in a zoo in this country. He lives in trees in the Malay Archipelago and is a most extraordinary little animal with huge eyes. Certainly he does resemble us more than the other lemurs do. His muzzle has been very much shortened, just as ours has, but perhaps that is not very significant, for so has the bull-dog's for that matter. A more important point is the socket for his eyes, which is nearly walled in with bone behind, and not open as it is in other lemurs. Then again its after-birth is a lumpy sort of thing, as with us, instead of having a membranous texture, as it has in other lemurs. But in both these points Tarsius resembles monkeys and apes just as much as it resembles man. Possibly the common ancestor of monkeys, apes, and man was an animal allied to Tarsius, but it seems very unlikely that we are more closely related to Tarsius than to apes.

The general conclusion which we have reached is that the recent physiological work, on uric acid and blood tests, confirms the older anatomical evidence that the apes are man's nearest relatives.

3. MISSING LINKS

WHEN I am dead, the chance that my bones will become fossilized is very remote. Bones decay away like the rest of our bodies unless a lot of very unlikely things happen. First of all, a dead body will not leave any permanent remains in the form of a fossil unless it happens to be covered up and thus protected from decay. That is fairly easy in the case of animals in the sea. Rivers are always carrying sediment out and depositing it, and tides and currents shift the sediment and cover up the bodies of dead animals. But even in this case it is by no means likely that the bones will be fossilized. Much more probably they will gradually dissolve away and leave no trace of themselves. Fossilization is rather a complicated process. It involves the replacement of each particle of bone, as it dissolves away, by a less soluble and therefore more permanent substance. When that has happened, the chances are still very remote that anyone will find the fossil thousands or millions of years later. Our quarries and mines and cuttings are mere scratches on the surface of the earth. With terrestrial animals the chances of fossilization are still less than with marine ones. They are likely to die and decay without being covered up. It would be quite absurd to look with any great hopefulness for the fossil remains of the ancestors of any given animal. It would not simply be like looking for the proverbial pin in a haystack, for then you are supposed to have the advantage of knowing that the pin is there. But in this case you are looking for a soluble pin in a haystack in a thunderstorm, and you always have at the back of your mind the disconcerting thought that perhaps it is no longer there.

That is the reason why we cannot describe the evolution of every species of animal in detail. The obvious thing to do is to study those animals which happen to have left the best record of their evolution. The horse is the best of all. We know the stages in the evolution of the horse in great detail, and with certainty. There are many other animals whose evolution from simpler forms is also well known. But if you take any animal at random, say a rabbit, the chances are that there will not be a complete fossil history of it.

One would not expect, then, to be able to find much in the way of human fossils, and the fact is that not many have been found. But we are in a very different position now from what we were at the beginning of the century.

At that time very little was known. A fossil skull had been found in a cave at Neanderthal in Prussia. This was definitely human, but had many ape-like characters. The enormous bony ridges above the eye are the most obvious features. Then there is the retreating forehead, receding chin, and massive jaw; and the form of the leg bones of this type of person shows that he must have shuffled along with his knees bent all the time. A cast of the inside of his skull gives a good idea of what his brain must have been like, and one can see from it that the parts of the brain concerned with speaking were poorly developed.

Now in the last century people did not like the idea of being descended from apes, and they were not prepared to examine the evidence for it impartially. They invented an excellent excuse for this skull. It was an abnormality! That would get out of the difficulty. The unfortunate individual had some disease which made his skull grow in that funny way. A little peculiar, was it not, that hundreds of thousands of his relatives, who of course had skulls exactly like ours, left no fossil remains, while just the single one who happened to be abnormal was fossilized! But improbabilities do not worry people who have convictions based on prejudice and not on love of truth. Some people even suggested that these skeletons were those of hybrids between men and apes. This is incredible for two reasons. Firstly, no cases are known of any two Mammals, so widely separated as to fall into different families, being able to interbreed. Secondly, even if one imagined the impossible, and supposed that such hybrids could be produced, it would remain incredible that the millions of normal men of those geological times should have left no trace whatever, while the few hybrids were by a miracle fossilized and discovered.

How has the famous Neanderthal man fared in our enlightened twentieth century? Many more skeletons have been found, closely resembling him. Neanderthal man has been found in Belgium, France, and Gibraltar, and in 1925 near the Sea of Galilee. With the skeletons are examples of his implements, which differ from those of other fossil men, and implements like these have recently been found in Mongolia. His was an enormously widespread race of primitive men, every one of them having those very characters which our learned and truth-loving forbears preferred to think of as due to disease.

In 1921 a fossil skull, without lower jaw, was found in Rhodesia. This had huge bony ridges above the eyebrows, and in most respects was rather like the Neanderthal man, but a little more primitive. We must hope for more examples of this race.

These Neanderthal men were fairly recent, as geological time goes, and also definitely more human than ape-like. They were probably not on the direct line of our ancestry, but died out perhaps twenty-five thousand years ago, just before the last ice age. Nevertheless they must have been closely allied to our ancestors.

Now what about the real missing link, something midway between ape and man? Where did we stand at the beginning of the century?

A most momentous discovery had recently been made. Dubois had set off to the East Indies with the avowed intention of finding a fossil ape-man, and, miracles of miracles, had actually found one in Java, after excavating for two years in Sumatra. It was sadly incomplete—just the top of a skull, a leg-bone and some teeth—but what was there was an amazing link between man and apes. If Neanderthal man's forehead may be said to recede, Java man's is almost non-existent, for his head slopes almost straight back behind his huge eyebrow ridges. His brain must have been about half-way in size between the brain of a gorilla and the brain of a man, yet he must have been about as tall as modern man. Here we have a very primitive man, or a very man-like ape, call it which you will, who existed—as the geology of the place shows—at about the time of our first ice age, perhaps half a million years ago.

That was rather a shock for the nineteenth century, and there was some attempt to discredit Dubois. Unfortunately for the disbelievers, however, the fossil bone was subjected to microscopical examination and proved beyond doubt to be genuine.

Since then there have been thrilling discoveries of intermediates between apes and men. I must pass over a lower jaw found near Heidelburg in Germany in 1907, although it is extremely interesting, simply because it is only a jaw. Four years later some workmen were digging gravel at Piltdown in Sussex, when a fossil human skull was discovered. This was a priceless specimen. One feels that one would have

K

sacrificed a hand or an eye to preserve this treasure so that it could be examined by an expert. What happened? Workmen, ignorant of its importance, broke it up and threw the pieces into a rubbish dump. By extreme good fortune Mr. Dawson had been on the look out for pre-human remains in the district for some time, as he had found peculiar flints among the gravel, and someone gave him one of the fragments. We must thank Providence for putting Mr. Dawson there, for he had the dump most carefully searched, and many of the fragments were found. Experts then set to work to consider how they should be fitted together, and different experts had different ideas.

The main conclusions are the following. There are scarcely any bony eyebrow ridges at all, and the forehead rises quite steeply above the eyes. This is most surprising in such an ancient skull, which is probably not very much more recent than the Java skull. But associated with this skull there was a lower jaw which is to all intents and purposes that of a chimpanzee. Many experts considered that it was an extinct chimpanzee's lower jaw. The complete absence of chin and the huge canine teeth supported that view. These canine teeth must have interlocked with those of the upper jaw like a dog's. Now if we regard the jaw as belonging to the skull, then we have a splendid missing link. But if they do not belong to one another, then the find is not nearly so significant.

That is why the recent discoveries near Peking are so tremendously important, for now an essentially ape-like lower jaw has been found in the same lump of rock as part of an essentially human brain-case, and the Piltdown skull and lower jaw are thus confirmed as belonging to one individual.

The story of the Peking discoveries is most interesting. During the war China started a geological survey, and got a Scandinavian, Dr. Andersson, to direct it. Dr. Andersson discovered rich fossil beds about forty miles from Peking. A great deal of excavating was done, but no human remains brought to light. One day one of the Chinese workmen was overheard asking a companion why they were wasting their time hunting for fossils in that particular place, when there were far more about half a mile away. That chance remark altered the course of our knowledge of man's ancestry, for the site of excavation was changed, and shortly afterwards human remains began to be found.

The first discoveries were two teeth, but there was nothing very special about these. Then in 1927 another tooth was discovered, which was sent to Dr. Davidson Black in Peking for examination. It was by no means by chance that Dr. Black was in Peking. Years before he had taken the Professorship of Anatomy at Peking, simply because he thought it likely that pre-human remains would be found in China, and he wanted above everything to carry out research on this subject.

Careful measurement of this tooth convinced Dr. Black that it was intermediate between a human and an ape's tooth. He exhibited the specimen widely, but it was received with scepticism.

A year later part of a jaw was found, and in the same piece of rock part of a skull. I have referred to that already. You will remember the jaw was essentially an ape's jaw, and the skull essentially human. Not only were these two bones found in the same block; they were both obviously of a young individual. There cannot be any doubt that they belong together, and they confirm the lesson taught by the Piltdown skull, that man retained the chinless condition of his ancestors till rather a late stage of evolution, when he had already got a large brain-case. Dr. Black was now enabled, by a grant from the Rockefeller Trustees, to devote full time to research. Discoveries were coming thick and fast, for in 1929 a momentous discovery was made by a Chinese geologist, Mr. Pei. Mr. Pei found an almost complete brain-case, quite uncrushed. Mr. Pei sent it to Dr. Black, and Dr. Black spent weeks in freeing it carefully from the rock in which it was embedded. Dr. Black has now described the skull, and casts of it have been made, one of which was exhibited by Sir Elliot Smith at the centenary meeting of the British Association.

Other finds have been made since. Altogether parts of about ten people have been found. The geological age of this primitive race must have been about the same as that of the Java man.

What are the essential features of the skull? Does it resemble Piltdown man closely? In one respect it certainly does not. There are large eyebrow ridges. The forehead is receding, and in this respect also it resembles Java man. In one way, however, it is like the Piltdown skull. If you put a finger on your head just above your ear, and move it across the top of your skull and down to the other ear, you will find that your skull is smoothly curved. This Peking skull is not smoothly curved like that. It has a distinct bump on each side opposite the part of the brain which is used for understanding spoken words, and another bump opposite the part concerned in using hand and eye together. This seems extremely significant. It looks as though man was just beginning to speak and use tools. As his brain swelled in the appropriate places, so his brain-case enlarged unevenly. This curious feature closely resembles one of the reconstructions of the Piltdown skull. Otherwise the brain was small, as we should expect in a missing link. Certain parts of the skull are very ape-like, especially the bones round the base of the ears, and of course the lower jaw was absolutely chinless and ape-like.

Let me summarize. Perhaps half a million years ago man was in a very ape-like condition, as shown by the Java, Piltdown, and Peking skulls. His brain-case was smaller, and his brain was just swelling in those regions which are concerned with speech and the use of tools. His skull was thick. His lower jaw was absolutely ape-like. These are the three missing-link skulls, though the term is, of course, no longer suitable. Then, ages later, we have a large number of skeletons and tools from various parts of Europe and Asia which belong to the Neanderthal type. This race much more closely resembles modern man. The chin is still small, though the lower jaw is by no means ape-like. The heavy overhanging eyebrow ridges and retreating forehead are persistent marks of the beast. Neanderthal man was probably fairly closely allied to a not very remote ancestor of ourselves.

You can find casts of some of the skulls and lower jaws to which I have referred in many museums. In the Natural History Museum in South Kensington they are in the room to the right as you enter. If you can find a skull of one of the aborigines of Australia in a museum anywhere, you will find it interesting to compare it with a European's, for it is primitive in many ways. Notice the small brain-case and the large eyebrow ridges and the receding forehead. The hairy Australian natives are the most primitive people living on the globe to-day.

Perhaps you will have come to the conclusion that scientists are apt to base a lot of speculation on very fragmentary evidence. The fossil skeletons I have mentioned are very incomplete, except the Neanderthal ones. As a matter of fact there is no undue speculation. Let me tell you a story which proves this.

A long time ago, when people were just starting to colonize New Zealand on a large scale, a colonist found a bit of bone in his garden. It was about eight inches long. The finder thought it might be interesting, and he sent it to Professor Owen in England. Professor Owen examined it carefully, and decided that it was a small fragment of a thigh bone of a huge unknown bird allied to the ostrich. He therefore published a paper saying that he supposed that there formerly existed in New Zealand a gigantic species of flightless bird, larger than the ostrich.

Now perhaps you think that he was basing too much speculation on too little evidence. But he was not. As New Zealand became better known, more and more bones were discovered, and now you can see whole skeletons of the great Moa of New Zealand in many museums. Professor Owen's speculation was proved to have been based on sufficient evidence.

4. THE EVOLUTION OF MIND

Just because we live such highly artificial, civilized lives, we almost begin to think that we have scarcely any of the primitive instincts which would have been of use to our ape-like ancestors. But deep inside us we still have those instincts, which often come to the surface in emergencies. You can easily prove to yourself that you have them.

Ask some friend whom you know to be a really bad driver to take you out in his car. Relax and allow your mind to wander where it will. Your friend decides to overtake another car round a blind corner. Here is a car coming straight towards you!

Now look at yourself. What are you doing? You are *holding on* firmly to something or other. It may be the door-handle, or the dash-board, or part of your seat, or even your friend's arm, or worse still the steering-wheel, but the fact remains that if your friend has really succeeded in frightening you, you will probably be holding on to something.

Now holding on is perfectly useless to you, and so it would be to any terrestrial animal. But think how absolutely essential it is to animals which live in trees. If they had not got a strong instinct to hold on when frightened, then they would often fall to the ground in emergencies and get killed. If our arboreal ancestor had not had that instinct, he would have fallen to the ground and got killed and you and I would not be here to-day.[1]

Here is another way in which you can show a primitive instinct at work, but it is rather more troublesome. If there is a large wood in your district, the middle of which is far from any house or road, go alone to the middle of it at about 1 a.m., and walk about a bit. You are very likely to find an almost overpowering instinct to return at once to the society of other human beings. It is not simply fear of the dark. People who are not in the least afraid of the dark in ordinary circumstances are afraid of walking about alone in woods at night. This instinct must have been of great importance to

[1] When this was broadcast, a listener (Mr. Worsley) wrote to say that when he was frightened in a motor-car, his toes contracted strongly, as though trying to grasp, and he was unable to uncurl them while the emergency lasted, although their contraction was painful. In my next talk I mentioned this, and many other listeners reported this instinctive curling up of the toes in emergencies, e.g., when being driven dangerously in motor-cars, when a passenger in a stunting aeroplane, in a ship in a violent storm, in difficulties on a precipice, when about to take an anaesthetic, and even when frightened by a description of an operation by a nurse. I have myself often experienced this phenomenon in motor-cars. It is experienced in both feet not only by people who drive but by those who do not, and is quite distinct from the automatic pressing of an imaginary brake by the right foot only, which those who are accustomed to drive often experience when being driven dangerously by someone else. It is very interesting that our instinct to grasp with the toes when frightened should have so long outlasted the capacity of the toes to grasp effectively.

our pre-human ancestors. The carnivorous enemies of early man would soon account for anyone who was foolish enough—or rather sufficiently lacking in this instinct—to walk alone in woods at night. Carnivores are far less likely to attack if two or more persons or animals are together, and this instinct completely fails to appear if two or more people are present. Our ancestors must have been especially easily attacked in woods at night compared with most animals. Their sense of smell, like ours, must have been poorly developed, so that they could not wind approaching enemies. Further, they must have been very defenceless before they took to using weapons, for their canine teeth had become reduced from their huge primitive dimensions. Their ability to escape by climbing was small in comparison with that of an ape with an opposable big toe. Escape by running would be difficult in a wood. No wonder their instinct not to wander alone in woods at night was strongly developed.

My brother, Mr. S. J. Baker, informed me that he found no such instinct while sitting in trees waiting for big game in Indian jungles at night, but that he experienced the strongest instinct not to descend. It seemed possible that this disinclination to descend might be rational and not instinctive in an Indian jungle, where there was actually more danger on the ground than in trees. I therefore decided to test the matter in a wood in England, where there could be no question of a rational fear, since there was no danger. In order to make sure that the instinct would appear, I got a friend to bring me into a wood which I had never been near before and leave me in it. I had no compass, no whistle, no stick, and no match nor light of any sort. I here quote my notes made directly after I left the wood. I have thought it best to present them exactly as I wrote them down. I was left alone in Whiteleaf Wood, near Monks Risborough, at 12.10 a.m.

"At first I stood still, and no instinct appeared. Then I began to walk, and found an instinct not to. However, I did walk, but instinctively very silently, putting my feet down slowly and as noiselessly as possible. I forced myself to crash along for half a dozen steps a few times while in the wood, but found it quite hard to force myself. When I made an unexpected noise in walking, I instinctively stood stock still, without moving any part of my body, even if I chanced to be in a peculiar position, as when stooping to pass beneath a bough. When I heard an unexplained noise not made by myself, I instinctively turned my head towards the source of the noise, or partly towards it, and then remained stock still, however peculiar my attitude chanced to be. I was constantly glancing back over my shoulders, and felt my back insecure. When I walked along quietly, giving all these instincts full play, I felt fairly comfortable, but I was extremely uncomfortable while forcing myself to crash along. Nevertheless I did feel an instinct not to walk at all, very strongly, and especially to remain still with my back to a big tree.

"I did not feel any instinct to climb a tree; but when I climbed one, I immediately felt perfectly happy and unconcerned, and did not care whether F. came back or not, except that I was afraid that I might fall if I went to sleep. I did not want to descend the tree and start walking again at all, but made myself do so. When on the ground again, I felt anxious to get back to the branches of the tree, although I did not feel like that until I had once experienced the sense of security that the branches of the tree gave me.

"I climbed the tree again, descended again, and climbed again, and renewed the sensations produced before.

"F. returned and brought me out of the wood. The feeling of security given by climbing was surprising to me. I was very sceptical about it before.

"I have written this down directly on my return to F.'s house. I must record that I felt just like what I imagine a wild animal feels like when walking about in the woods—alert, suspicious, on the *qui vive* all the time except when in the tree."

Anyone who doubts the reality of these feelings should repeat this experiment. It is essential that he should be in a strange wood, with no way of finding his way out, and he must be quite alone, and it must be the middle of the night.

It appears to me quite possible that we are only terrestrial by tradition and not by instinct. Our instincts may still be arboreal, as in many children. In exactly the same way the otter is not instinctively aquatic. Far from it. Every young otter must be forced by its mother to enter the water against its will. It must be taught by its mother to swim. I could give you many instances of the importance of tradition in animals. The wild children who have occasionally been found have usually been arboreal, and their extreme agility in trees has made it difficult to catch them. It should be recalled, too, that most people prefer to go upstairs to bed.

Perhaps this consideration of primitive instincts will have paved the way for a comparison of the brain of man and apes. It is important to remember that we may not be so superior to the ape as we seem, for this reason. Much of our apparent cleverness is due simply to the ability of our minds to comprehend what others have discovered and communicated to us by speech and writing. Speech and writing tend to give us a very conceited impression of our own brains, for they enable us to profit from the wisdom of the ages in a way which would be totally impossible to the chimpanzee and orang-utan, even if, except in the matter of speech, he had the same innate mental powers as we have. There is no evidence that the innate intelligence of man has increased in the slightest degree during the historical period. Discoveries have been made, and speech and writing have enabled these to be broadcast, so that each generation piles on new knowledge and discards what it proves to be erroneous. Thus knowledge increases, but it seems certain that the brain is not evolving.

I think there can be no doubt that articulate speech separates us more from animals than anything else. Without speech, what knowledge of the universe should we have to-day? Fancy yourself cast up on an uninhabited island as a child before you had learnt to speak. Suppose that it was a land flowing with milk and honey, so that you did not simply die at once. How much would you find out about the universe before you died? One cannot say, but it would be very little. Do you think you would have been able to distinguish six objects from seven? Certainly it seems unlikely that you would have got as far in the multiplication table as twice two make four. When we think of the great geniuses of history, we must appreciate the great extent to which they relied on the world's store of knowledge existing at their time in the form of speech, whether spoken or written. Speech, and especially written speech, enables us to start where our ancestors left off. The elementary student of biology to-day knows more about evolution than Charles Darwin ever did.

Can any apes speak at all? It all depends on what we mean by speech. The gibbon and the chimpanzee certainly have vocabularies. Definite sounds, which we can reproduce by phonetic spelling, have definite meanings. But none of these sounds indicate anything except emotional states. The chimpanzee can say something meaning "extremely pleased," or "very fond of you," or "bored," or "hostile." But he has no name for any concrete object, not even for a banana, or a tree, or water. The evolution of speech in our sense must have started by some of our pre-human ancestors beginning to attach definite sounds to definite concrete objects. The great advantages accruing from even a rude form of speech would result in the natural selection of the speakers and their offspring in the struggle for existence.

Apart from speech, is there such a tremendous gap after all between a man's mind and an ape's? It is incredibly hard to know what is going on in the mind of an animal. When I see a cow, I often think to myself what an extraordinary blank in knowledge its mind represents. I know its anatomy perfectly well and a certain amount about its physiology. The world's store of knowledge on these subjects is immense and almost incredibly detailed. But its mind! When it lies there chewing the cud, I have not the very slightest idea what it is thinking about, or if it is thinking at all. Is it turning over the events of the day, or wondering about the future, or is its mind an almost impenetrable fog, like our own when we are half asleep? Has the cow consciousness at all? This question does not seem at all ridiculous when we consider what complicated things human beings can do without consciousness when they are sleep-walking.

Animal-lovers step in where scientists almost fear to tread, and they have produced amazing and quite incredible stories of the sagacity of the dog. These stories are quite useless to anyone who really wants to explore the mind of animals. They are recorded by wholly uncritical people who are quite unable to view a dog's behaviour objectively, but are carried away by emotion and love of the remarkable. I should be the last person to minimize the intelligence of the dog, for I have always lived with dogs, and I am very sensible of the astounding capacity they have of interpreting inflexions of the human voice, and also of recognizing voices. One of my dogs never paid the slightest attention to the wireless until he heard me broadcast. The moment I started to speak he recognized the voice, and paid attention to the wireless for the first time. But such incidents as these really lead nowhere. They are not much more helpful than the obviously untrue dog-stories. There once existed a club in Oxford whose function it was to invent dog-stories to send to a weekly paper. A good deal of shrewdness must have been needed to gauge the editor's credulity.

After stories of this kind, a little objective research

into the mind of animals comes like a breath of fresh air into an ill-ventilated room. Professor Köhler, of the University of Berlin, has made such an objective research into the mind of chimpanzees and has described it in his book *The Mentality of Apes*. While nearly the whole civilized world was proving its civilization by mutual destruction in the world-war, he was carrying out his really thrilling investigations in the island of Teneriffe. Professor Köhler had no preconceptions, no desire to make out that his chimpanzees were more or less intelligent than they were. His idea was simply to test their intelligence in an entirely impartial way.

The tests he devised were excellently thought out and quite different from the usual tests. Let me tell you what he was *not* doing. He was not testing their ability to imitate human beings, or to learn to perform complicated actions. He was not testing them as so many people have tested animals, who think they are testing intelligence when really they are doing nothing of the sort. A favourite method with these others has been to put the animal in a cage, from which it can only escape by pressing a button which releases a door. In more complicated tests the animal has to move several buttons and latches in a certain order. The mistake in all these tests is that the mechanism of the action of the button in releasing the door is invisible, and though the animal soon learns the trick, he can have no insight whatever into its mechanism. People who have experimented in this sort of way have often concluded that animals are wholly lacking in intelligence, but really it would be as sensible to say that a man was wholly without musical taste, when one had only tested his capacity as a weight-lifter.

What, then, was the essence of Professor Köhler's experiments? Simply this, that in every experiment the whole solution of the problem set should be clearly visible to the chimpanzee.

The experiments were performed by putting food where a hungry chimpanzee could see it, but could only get it if it exercised intelligence. The simplest test of all is to put a banana outside the bars of the cage with a string attached to it. All the chimpanzees tested at once pulled the string and obtained the fruit. Dogs are generally non-plussed by this, though it would be simple for them to pull the string with their teeth, if they had the intelligence to comprehend the situation.

If a stick lies in the cage and a banana is placed outside, out of reach, the behaviour of the chimpanzee depends on where the stick is placed. If it is placed in such a position that it can view the banana and the stick at the same time, it uses the stick as an implement to drag the food towards the cage. If, however, it cannot see the stick at the same time as the banana, it seldom has the sense to use it, though it may look straight at the stick from time to time.

In solving these and other problems it is clear that

the chimpanzee has a real grasp of the situation. It is not that it behaves at random in the first instance till by luck it secures the fruit, and that afterwards it repeats identically the same movements of the same muscles as gave success before. That is what some observers would have us believe. On the contrary, the ape acts stupidly for a time, and then suddenly grasps the situation. From that moment its action is sure and decided, in marked contrast to its undecided movements a moment before. Next time the same problem is set, it solves it more quickly, but often by quite different movements of its body. Success comes from mental grasp of the situation, and not from repetition of certain bodily movements which chanced to bring success before.

Curiously enough, chimpanzees are extraordinarily stupid about removing obstacles, though quite sensible about using tools. If a box is placed inside the cage in such a position as to prevent the ape from getting into a position from which it could reach a fruit placed outside, only the most intelligent of the chimpanzees have the sense to move the box; and even they take a long time to see the obvious solution of the difficulty.

The chimpanzees become accustomed to obtaining fruit placed high up, out of reach, by swinging on a rope attached to a horizontal beam. This gave an idea for an excellent experiment. The rope, instead of hanging free, was wound round the horizontal beam in three neat loops, not crossing one another. It now became apparent that three neat loops appear to a chimpanzee precisely as a hopeless tangle appears to us. They all tried to untangle the rope, but quite unmethodically, just as we often try to straighten a tangle of string quite unmethodically. If our brains were much more efficient than they are, we should straighten tangles with sure, decided movements, never making the tangle more complex. Before judging the chimpanzee too harshly, however, we must remember, as Professor Köhler remarks, that many of us find a deck-chair as inextricable a tangle as a chimpanzee finds three loops of rope.

What is the limit of chimpanzee intelligence? I think that this is the hardest test that any of Professor Köhler's animals passed. The fruit is placed out of reach outside the bars of the chimpanzee's cage. A stick is provided, but it is hung on the wall so high up that it can only be reached by dragging a box below it and climbing on it. It takes a very clever chimpanzee to see what to do.

Professor Köhler considers that the chimpanzee may actually be nearer to man in intelligence than to many of the lower species of monkeys.

I cannot here discuss the expression of the emotions in animals. Charles Darwin wrote a fascinating book on this subject. On the whole the chimpanzee expresses his feelings very much as we do, even to scratching his head when he is puzzled and beckoning with his finger when he wishes his friend to approach.

Next time you see a chimpanzee in a cage, you will probably not view him with quite such contemptuous amusement as before. One should think how one would behave if one were shut up in a cage oneself, with no privacy and no clothes, let us say in Japan, or in some other country where one could not make oneself understood. Of course clothes give us a wholly artificial feeling of superiority. Someone has questioned whether we should have much respect for the House of Commons if members were compelled to sit in a state of complete nudity. Even the House of Lords would lose something of its dignity if subject to the same restriction. Yet our respect should undoubtedly be for their brains and disinterestedness, and not for their clothing.

So far we have only discussed the changes in our outlook on the relationship between man and apes. Perhaps you will ask this, "Is that the only way in which the biological outlook has changed since the end of last century as far as the mind of man is concerned?" It is not, for many of our new ideas apply equally to man and to animals, profoundly affecting the way in which we regard man; so now we must take a more general outlook on modern developments in the matter of the mind.

Probably you know what conditioned reflexes are. In case you do not, I may first of all remind you what ordinary reflexes are. If I were to touch your hand with a red-hot poker a message would go along your sensory nerves to your spinal cord and another message would come back along your motor nerves to the muscles moving your arm. Your whole arm would be pulled away with great speed, so quickly that the whole movement would be finished before the fact that you had been burnt had arrived at your consciousness. Undoubtedly a great number of our actions are reflex acts in which consciousness is not involved. Now what are conditioned reflexes? If I were to play the note middle C on the piano a great number of times, presumably it would not have any special effect upon you beyond being rather boring. But if I were to prick your finger with a needle every time I played the note, you would develop a reflex action of quickly drawing your hand away every time the note was struck, even though I no longer pricked you. This is a simple example of a conditioned reflex. Another example will make the subject even more simple.

When one is hungry, the smell of food causes a reflex secretion by the salivary glands, so that one's mouth waters. That is a simple reflex action. There is no question of voluntary action here, because you cannot make your salivary glands secrete by any act of your will.[1] But conditioned reflexes often grow up round these reflexes. If a certain gong is used to summon you to your meals, just the mere

[1] A dentist has written to tell me of a woman patient who can open her mouth and eject a fountain of saliva at will. Presumably she uses a voluntary muscle to compress the salivary gland. There is no evidence that she *secretes* saliva at will.

sound of that gong will cause increased secretion by the salivary glands. Some very interesting experiments on this subject have been done with dogs. The conditioned reflex of watering in the mouth at the sound of a bell is soon developed if dogs are always fed immediately after a bell is rung. There is no difficulty in showing that; because if you ring a bell but produce no food, saliva pours out into the dog's mouth. That is simple, but here is something much more interesting and unexpected. If you always ring a bell a quarter of an hour before you feed your dog, then your dog will develop a conditioned reflex of secreting saliva a quarter of an hour after the ringing of that bell.

It seems likely that conditioned reflexes play a large part in our ordinary behaviour. Many of our actions may be attributed to them, and this is an important and expanding field of research at the present moment. Nevertheless, it is possible that some enthusiasts have gone too far in regarding animal or human behaviour as almost exclusively made up of an infinitely complicated system of conditioned reflexes.

Then there are the ductless glands. So much research has been done on them lately, and the results obtained have been so exciting, that they have found their way into the popular newspapers, despite the fact that their study is a branch of science. What are the most important conclusions that we can draw from this study, so far as the mind is concerned? That is fairly easy to answer.

We now see that many of our actions and emotions are by no means wholly to be ascribed to our nervous system. There exist in our body definite glands which pour definite chemical substances into our blood-stream, which profoundly affect our thoughts and our actions. The most important are the thyroid gland near the Adam's Apple in the neck, the pituitary gland between the roof of the mouth and the brain, the adrenal glands near the kidneys, and the glandular tissue of the reproductive organs.

What are the effects of these glands? The straightforward way to find out is to remove them in the case of animals, and see what happens; or else we can notice what happens when they are diseased. Experiments and observations of this type have conclusively proved that our emotions are very largely controlled by chemical substances circulating in the blood which have been produced by the glands. It is interesting to note that the chemical substances formed in each gland are the same in all mammals which have been investigated, including man. It has been possible in the case of some of the glands to isolate the actual chemical compound, and to analyse it, and in some cases to synthesize it in the laboratory.

The thyroid is the easiest gland to take first. Its secretion has the effect of increasing the rate at which bodily processes take place, and not only bodily but also mental processes. The person whose thyroid gland is not functioning properly is not only sluggish in his movements, but also in his brain. He cannot help it. He must be provided with more of the essential substance secreted by the thyroid gland, and then he recovers. That is quite simple, because the substance is easily extracted from the thyroid gland of cattle or sheep killed for food. Sometimes the thyroid gland of young children is almost wholly deficient, and then cretinism results. In extreme cases the cretin may have the bodily appearance and the mental capacity of a child of two, when its actual age is ten times as great. This condition can be relieved by taking tablets of thyroid gland by the mouth.

Some people have not too little, but too much thyroid gland, which may succeed in secreting too much of the essential substance into the blood-stream. This results not only in undue physical activity and bodily restlessness, but also in an unduly active mind and mental restlessness. Such people are usually thin and have an agitated expression. The eyes often protrude from their sockets.

One must not fail to mention the adrenals. If you inject some of the secretion of the adrenal into a man, all the symptoms of fear are produced. The face goes white, the hair stands on end, the blood-pressure is increased, the heart beats strongly so that its palpitations are easily felt, and the normal movements of the small intestine are arrested.

In the case of the reproductive organs, the matter is simple, for the development of the sex instinct depends on substances produced by the reproductive glands. If they are removed in early life, these instincts never develop. By removing the reproductive organs from an animal, and grafting instead those of the opposite sex, one reverses its sex instincts. In this way one may cause a male guinea-pig not only to produce milk, but also to acquire the instinct to suckle young.

There is no doubt that the mechanism is the same in man, and the utmost care should be taken in cases of abnormal sexual behaviour, to make sure that the glands are normal; for if one were to punish a person for unusual behaviour when the glands were abnormal, one would be doing a grave injustice. You or I would behave abnormally if the inappropriate glands were grafted into us.

The fact of our relationship with animals is well shown by recent experiments performed in France. It was necessary to remove the ovary of a woman by an operation. A fluid was removed from the ovary, and injected into some ancient female rats which were so old that the sexual instinct had been lost. The result of the injection was that the instinct was immediately reacquired.

It cannot be denied that the work on conditioned reflexes and on ductless glands has given us a more mechanistic conception of the mind. It seems much more probable now that the mind is a mechanism whose physical basis at least may one day be interpreted in terms of physics and chemistry. I have only just been able to touch the very fringe of this subject. I wish that I could give you some idea of the enormous amount of work that has been done. The investigators have performed experiments and made observations which are repeatable. Anyone who doubts them can experiment for himself. The more one studies, the more one is absolutely forced to the conclusion that chemistry plays a large part in the control of our emotions.

This raises the old problem of mechanism and vitalism. Are the body and mind to be regarded as a machine working simply according to the laws of physics and chemistry, or is there some fundamental distinction between animate and inanimate matter? It seems to me that to be a vitalist is to fight a losing battle all the way. A hundred years ago, the first organic substance was synthesized from inorganic materials in the laboratory. Previously it had been thought that such synthesis was impossible. This was the first blow to vitalism. No other organic substance was synthesized for some time, and the vitalists became optimistic that perhaps this was the one exception that proved the rule. But since then many such substances have been synthesized from inanimate substances in the laboratory, and to-day no one would claim that any particular substance in the body will *never* be synthesized. The vitalist simply says that he thinks that we shall be unable to interpret life in terms of physics and chemistry. But the trouble is that every discovery in biology brings us nearer to such an interpretation. Every discovery makes his position less tenable.

Are we then to pronounce ourselves as mechanists? That, you will say, is the obvious alternative. Personally, I consider that it would be premature in the extreme to do so. We must only be mechanists when all bodily and mental processes have been reduced to mechanics, physics, and chemistry, when we understand precisely what chemical and physical changes underlie and constitute every part of every action, every thought, every memory, every emotion, even consciousness itself. As yet we do not begin to approach that position, and I consider that the only reasonable position to take up at the moment is that of scepticism. Nevertheless, let us remember that at present all discoveries in biology are leading us nearer and nearer to a mechanistic explanation.

The recent discovery that excessively small particles of matter do not obey the ordinary laws of physics, which are only applicable to large masses, has been interpreted by some as making probable the existence of free-will. I agree with Professor Levy that any unprejudiced person, if he really thinks seriously on the matter, will agree that there is no connection between the two subjects.

5. THE CONTROL OF DEVELOPMENT

I HAVE mentioned various changes in the way in which we regard man, which have come about during the course of the present century. You will have noticed that the new work in anatomy and physiology is a logical outcome of the old. There have been thrilling discoveries, but it cannot be said that we look at man in a fundamentally different way as a result of those which we have discussed so far. That does not apply at all to what I shall now relate, for in the realms of inheritance and sex determination we now know much, where practically nothing was known before. The beginning of this century was the time when the new knowledge suddenly began to spring into existence.

In a limited sense that is not quite true, because thirty-five years earlier the monk Gregor Mendel had made his momentous discoveries on inheritance by experiments on peas in his monastery garden. But these discoveries were absolutely unknown to the world at large, and their rediscovery, together with the tremendous outburst of research which they stimulated, only took place about the beginning of the new century. The world's knowledge of inheritance before was almost negligible. Mendel's laws were found to have universal application not only in plants and lower animals, but in the higher animals and in man himself.

The 21st of July, 1901, is a significant date in the history of biology and of civilization. One day, perhaps, children in schools will no longer learn the dates of the accession of kings, but of really important happenings in art, music, and science. On that date the American investigator, C. E. McClung, sent a scientific paper to the German periodical *Anatomischer Anzeiger*. In that paper he put forward his theory of the determination of sex by chromosomes, which has been substantiated by all subsequent research.

Now, how had McClung made these discoveries which will result in his name being honoured centuries hence when many ephemeral notorieties of to-day are known by no one? You will probably laugh. It was something so peculiar that it would only be natural if it were to cause mirth. It is a splendid illustration of the fact that those who do very peculiar and eccentric things are commonly those who do important things. If one has a child with peculiar and apparently useless traits, it may be better to allow him to develop them than to attempt to mould him to an ordinary pattern.

What had McClung been doing? He had been studying grasshoppers. That in itself appears to have been rather a useless and eccentric occupation, yet when the day comes, as it surely will, when we are able to control the sex of our offspring and of our domestic animals, it will be primarily due to

McClung's studies on the reproductive organs of grasshoppers.

The proper study of mankind is by no means always man. The proper study in this case would have been the reproductive organs of grasshoppers, simply because for technical reasons the conditions are easily studied in them. It must suffice to say that the control of sex could never have been discovered by the investigation of the reproductive organs of man. Since the time of McClung's first announcement, an enormous amount of research has been done on this subject in the most widely diverse groups of animals up to and including man; and as man is our chief interest here, we must now concentrate upon the determination of sex in man. What determines whether a given embryo is male or female?

First of all, every woman produces one egg every month. That egg has no tendency to grow into an embryo of one sex rather than the other. With men it is different. The reproductive cell is here a microscopic structure shaped roughly like a tadpole and progressing in the same way by movements of its tail. Now these little tadpole-like sperms are of two sorts, male-producing and female-producing. The egg is fertilized by only one sperm. Should that one chance to be a male-producing one, a boy will be conceived. Should it be a female-producing one, a girl will be conceived. The determination of sex rests entirely with the sperm. Then, you will say at once, what decides whether a given sperm shall be male-producing or female-producing? To that there is a complete answer.

Very probably you have already heard of chromosomes. They have often been discussed in broadcast talks. They are minute, microscopic, generally rod-shaped bodies discovered about 1880. A sperm is extremely small, but chromosomes are much smaller, for the head of the sperm of man contains twenty-four of them, never more nor less, but always exactly twenty-four. You can easily imagine twenty-four rods of varying lengths, some so short as scarcely to be rods at all, some long enough in proportion to their breadth to resemble an ordinary ruler viewed from the flat side. Twenty-three of these chromosomes do not interest us for the moment, but the twenty-fourth does. The twenty-three are the same in every sperm, but the twenty-fourth is not. Either it is a very big one, one of the biggest, or it is a small chromosome, among the smallest. There are two types of sperms, in exactly equal numbers. One type has the big one, and the other the small one. These special chromosomes are called the sex-chromosomes, because they decide the sex of the embryo. The sperms containing the large sex-chromosome give rise to girls, and the sperms containing the small sex-chromosome give rise to boys.

The egg which is waiting to be fertilized by a sperm also contains twenty-four chromosomes.

Twenty-three of them are exact pairs for the twenty-three chromosomes of the sperm. The egg also contains a sex-chromosome, but it is always the same, always a large sex-chromosome, an exact pair for the large sex-chromosome of the female-producing sperm.

When fertilization takes place, that is, when a sperm fuses with an egg, forty-eight chromosomes are brought together. Forty-six of these are ordinary chromosomes, and the other two are the sex-chromosomes. If the two sex-chromosomes are *both* large, the embryo will grow up to be a girl. If one is large and one is small, it will grow up to be a boy. Its sex depends simply on its sex-chromosomes. The fertilized egg-cell divides into 2, the 2 into 4, the 4 into 8, the 8 into 16, and so on, until all the millions of microscopic cells are formed which constitute your body. Each time a cell divides, each of the forty-eight chromosomes divides. So if you are a woman or a girl, you have two large sex-chromosomes in every cell in your body, in the cells of your skin, your brain, your digestive organs, in every part of you. If you are a man or a boy, you have one large and one small sex-chromosome in every cell in your body. It was because of that that you grew to be a man. If the egg from which you grew had been fertilized by a sperm containing a large sex-chromosome, you would have been a girl.

I have said that the two sorts of sperms are formed in exactly equal numbers, so no doubt you

M

will at once ask, "Then why are not exactly the same number of boys born as girls?" Actually about 105 boys are born to every 100 girls in most European countries, and a considerably higher proportion of boys even than that are conceived. The male embryo is much more likely to die than the female, and if that were not so, there would be a great preponderance of boys. Males tend to die off at all ages more than females, and so there is a preponderance of females despite the birth of more males.

How can we account for the conception of more male than female embryos, if the male-producing and female-producing sperms are produced in exactly equal numbers? It seems that the male-producing sperms either swim faster, or else have some other advantage which enables them to effect fertilization more easily. The sperm has a big journey in front of it, when one considers its minute size. It has to swim nearly two thousand times its own length up the cavity of the womb. If it were magnified to the length of ourselves, its journey would be more than two miles.

These remarks about the size of sperms are a digression, but the subject is so interesting that I propose to digress further. I believe that someone has calculated the total weight of all the sperms fertilizing all the eggs which grow into babies in the United States in a year. Not being able to obtain these figures, I asked my brother, Mr. S. J. Baker, to make similar calculations. The results are so surprising as to be almost unbelievable. If you took all the sperms fertilizing eggs resulting in the birth of babies during a thousand years in the whole of England and Wales at the present birth-rate, they would weigh about as much as a pin's head. That is the total weight of the male contribution to the continuance of the English and Welsh peoples during a thousand years!

Another way of getting an idea of the minute size of sperms is this. An enormous number of sperms always compete for the fertilization of a single egg, yet only one succeeds. I calculate that one ordinary man produces in a year about enough sperms to achieve the conception of as many infants as are born in England and Wales in seven hundred years, at the present birth-rate, if every sperm were to achieve fertilization.

In the chromosomes of these microscopic sperms are borne the factors which react with those of the chromosomes of the egg and with the environment to make us what we are, to determine the structure of every part of the body.

The sex-chromosomes do not only control sex. They also carry the factors for many other characters. Take colour-blindness. The large sex-chromosome carries *either* the factor for colour-blindness *or* the factor for normal colour-vision. A single large sex-chromosome never carries both, but a woman has two large sex-chromosomes, as I have mentioned, and so she may have the factor for normal vision on one and colour-blindness on the other. In that case, luckily, the normal one over-rides the abnormal, and she does not show the least trace of colour-blindness. Her sons, however, only have one large sex-chromosome, and it is derived from her. Half her sons, on the average, will get the normal sex-chromosome from her, and half will get the one bearing the factor for colour-blindness. So, on the average, half her sons will be colour-blind. If she marries a man who is not colour-blind himself, all her daughters will be normal. She can only produce colour-blind daughters by marrying a colour-blind man. Half her daughters will then be colour-blind, as both their sex-chromosomes will bear factors for colour-blindness. That is why men are so much more commonly colour-blind than women. Colour-blind men can be produced when both the parents are apparently normal, but colour-blind women cannot be produced unless a colour-blind man marries a woman who is herself colour-blind, or at least has it in her family.

There are several other diseases which are inherited in exactly the same way, and of course with these also man is the chief sufferer. Haemophilia is a good example. That is the disease in which one bleeds profusely from a small cut. It is commonly supposed to be due to the skin being malformed, but that is not so. It is caused simply by a failure of the blood to clot.

One must not think of the large sex-chromosome as being concerned only in the control of sex and the inheritance of colour-blindness and haemophilia and other diseases. That is not so at all. Boys generally resemble their mothers rather more closely than they do their fathers. This is probably because they get their large sex-chromosome from their mothers, with all the factors it bears controlling development. The small sex-chromosome, which comes from the father, is quite inert and bears no factors.

The chromosomes other than the sex-chromosomes also carry factors controlling development. I have remarked that it was in 1901 that McClung first suggested that the chromosomes determined sex. It was about a year later, on October 17th, 1902, that another American, W. S. Sutton, sent a paper to the *Biological Bulletin,* in which he pointed out the close correspondence between what chromosomes do and what the factors of Mendelian inheritance do. Nowadays practically all biologists agree that the chromosomes are responsible for Mendelian inheritance. Like McClung, Sutton arrived at his conclusions from a study of grasshoppers.

There are many simple cases of Mendelian inheritance in man. If a girl with pure blue eyes marries a man with brown eyes, who received a factor for brown eyes from both his parents, then all the children will have brown eyes. These children will really be hybrids for this character, but brown eyes completely over-ride blue, and no trace of the blue appears. The fact that they are hybrids for this character appears when they grow up and marry. If one of them marries a person with pure blue eyes, the children will not all have brown eyes, as one might expect. On the contrary, half the children will have blue eyes and half will have brown eyes. Of course in a modern small family one might not get equal numbers, but the larger the family, the closer the approximation is likely to be; and if one took all such families in Great Britain, the approximation to the ratio of one blue to one brown would be very close indeed.

Most parts of the body do not present such simple cases of Mendelian inheritance, because several pairs of factors often affect each part. Thus at least four sets of factors are concerned in the shape of the nose, and therefore the results are not nearly so clear-cut nor so simply explained.

I want to make it clear that though you may have brown eyes yourself, and though your cousin may also have brown eyes, yet you are quite likely to produce some blue-eyed children if you marry her or him. Your cousin and yourself have a common grandfather and grandmother. Now your grandparents themselves may have had brown eyes, but one of them may have been hybrid for it, and transmitted a factor for blue eyes both to your father and to your cousin's father, and thus to you and your cousin, though none of these members of the family have had blue eyes. Now suppose you marry your cousin, some of your children may have blue eyes simply because blue eyes come to them from both parents.

Now blue eyes are of course most desirable characters in the eyes of most of us, but some undesirable characters are inherited in the same way. One of these is congenital feeble-mindedness. I hope I have made it clear that *feeble-mindedness* might result from a cousin-marriage, though there was no history of mental defect in the family.

This brings us to the whole question of inbreeding, about which our knowledge has increased so much recently. There is nothing whatever harmful about close inbreeding in itself, but it does very quickly bring to light any latent undesirable inherited qualities. If the stock is completely free from these, inbreeding has no bad effect whatever. The Pharaohs used to marry their sisters generation after generation, and experiments on animals have proved that once one has got the stock free from factors for undesirable qualities, close inbreeding is harmless.

When two congenital mental defectives marry, all their offspring are likely to be mental defectives. If a mental defective marries a normal person, all the offspring are likely to be normal, but to transmit mental defect. That brings us to Eugenics, and now for a moment we must leave pure science and become applied scientists. In pure science there are no values.

The pure scientist simply describes. He does not say what is good or bad. Nevertheless, I cannot help feeling that mental defect is fundamentally bad, and to allow congenital mental defectives to produce children seems sheer madness. Actually they are increasing in numbers in this country at a frightening rate. Unless something is done about it, we shall before very long find ourselves in grave danger, as they are among the most fertile people, while the more desirable stocks are failing to reproduce themselves.

Many kindly people have been turned against Eugenics because they have imagined that Eugenists are snobs or anti-humanitarians. I hope very much that I may be able to controvert that view. The life of a mental defective, or of a person suffering from some severe inherited disease, is one long misery, and it would be humane to prevent such individuals from being born. As to snobbishness, the real Eugenist is the last person to be a snob. He wants everyone to have an equal chance in the world, so that the inherently best people, from whatever class, may be given a chance of showing their desirable qualities. He must then try to find means of encouraging their reproduction.

If a rich person wanted to leave his wealth for the purpose of decreasing the amount of suffering in this world, I do believe he could scarcely do better than leave it for the cause of Eugenics. It may be true that snobs and anti-humanitarians have in the past been drawn to the subject, but they have been drawn mistakenly. The real Eugenist sees a practical way of decreasing suffering and spreading wider the desirable qualities of our race and other races. Let me give you an example of what could be done.

Quite recently a Bill was brought forward in the House of Commons to legalize Eugenic sterilization. Had the Bill become law, it would have begun a new era in the treatment of mental defectives. It must be understood that the sterilizing operation is in no sense a maiming operation. It does not make one impotent, but only sterile. The Bill provided adequate safeguards. No person was to be sterilized unless suffering from inherited mental defect, and unless his own consent was obtained (if capable of comprehending the matter), as well as the consent of the parent or guardian or wife or husband, and also of the Board of Control. It had also to be made certain that no danger to health was involved. Despite all these safeguards, the House of Commons refused leave to introduce the Bill.

Nevertheless it seems certain that the Eugenic conscience of the nation will eventually be aroused.

JULIAN HUXLEY

6. MAN AND REALITY

In the first contribution to this series I tried to give a picture of man as a relative being—to show how his construction, his way of working, and even his way of thinking, are only comprehensible in relation to his environment. I shall now attempt something, I fear, rather too ambitious—I shall attempt to show some of the ways in which the changed picture of life given by modern biology is helping to determine a change in the general picture of the world which we can draw for ourselves. For you must remember that man's general picture of his world evolves just as much as man himself. Indeed, it must evolve—it cannot stay still: the very notion of fixity of dogma, or of knowledge, or of ideas is an error and is wrong. One of the greatest marks of the modern world is the realization that while truth *must* always be incomplete, it yet can be progressive.

In recent years, several men distinguished in physical science—notably Eddington and Jeans—have given us a lucid account of their world-picture. Like all modern world-pictures, it has a basis in the facts and theories of science, and extends over into philosophy. I will not presume here to expound their views, which have after all been published and widely discussed. But there is one point about them by which a biologist cannot help being struck. The picture which emerges is of an observer contemplating a world from the outside, as Jupiter was supposed to contemplate Earth from the vantage-point of heaven. The observer is not so much an individual human being—Sir James Jeans or Sir Arthur Eddington personally—but rather an artificial creation—the human intellect, or perhaps better the human intellect as represented by the highest achievements of mathematical physics. And the world observed is not so much the concrete world as the ordinary man observes it, but a world with all its qualities taken out of it except what can be weighed and measured—a world of mass, space, time, and energy, a metrical, mathematical world.

That is all very well. It may be the simplest method of approach for the physicist. He is so accustomed to thinking in terms of inanimate things to be measured and reasoned about by his inquiring mind that he has come to take the mind for granted as somehow outside the things with which he as a scientist is concerned. But this is not necessarily the only or even the best way. The biologist, for instance, cannot see the world in this way. He has a more difficult job, for he knows that his mind is just as much a part of the things which he as a scientist has to investigate as is his body or the lifeless world around him. He must try to account for the observer as well as for what is observed: he cannot be content to leave mind outside the field of his facts, but must make it, too, part of the subject-matter of his science.

To understand the situation properly, we must look back a moment at the history of thought. What I am going to say about the so-called primary or secondary qualities of things is rather unfamiliar or perhaps difficult; but it is of the greatest importance for understanding the change which has recently come over scientific philosophy. Children, savages, and primitive philosophers are alike in believing that the qualities of objects are somehow *in* the objects. They would say that in an orange, for instance, there inhere the qualities of being round, yellow, having a certain pleasant taste, of weighing so much, and being of such and such a size. But with the rise of the scientific method in the seventeenth century, people began to distinguish between different kinds of qualities of objects. Some of them are still thought of as belonging to the objects themselves: these were called primary. The fundamental primary qualities were mass—which they defined as the amount of weighable matter in a thing; and magnitude—its size and shape. But other qualities, it appeared, were put into objects, so to speak, by us; and such were called secondary qualities. Colour is an example, and so are taste and smell. An ordinary man sees an orange as yellow: but to someone who is totally colour-blind it simply looks grey. And of course even before this, common sense as well as science had learned that whenever men ascribed emotion or will or purpose to a lifeless object, they were just projecting their own feelings into it. Savages think of the thunder as a manifestation of anger, because they are frightened by it. Early religions put benignity and power and wisdom into their idea of the Sun, because the Sun warms man, ripens his crops, and looks down upon the earth.

Thus first of all will and emotion were taken out of objects, and then all the secondary qualities, so that nothing was left but what you could weigh and measure directly.

But to-day we are reaching a new stage. For one thing, we are seeing that the so-called primary qualities are not any more in the objects than are the secondary. They are just the most convenient ways—most convenient, that is, from the point of view of science—of describing how objects appear to us. They are the most convenient because they can easily be measured in terms of quantities which all human beings agree about. Furthermore, other qualities can be referred back, so to speak, to these standards; colour, for instance, can be referred back to light-waves of a particular speed and size hitting a special kind of chemical substance in the retina of our eyes.

But they need not be thought of as *in* the objects. So far as we can get a picture of physical reality, the world is not made up of bits of matter of a definite size and mass, moving at definite speeds, like innumerable tiny billiard-balls careering through space. That was the first crude idea of the picture revealed by modern physics with its discovery of atoms and electrons. But now we are beginning to realize that instead of the ultimate units of matter being like our idea of ordinary material objects, only much smaller, they are something wholly different. They are centres of energy, whose effects shade off into remotest space. As Whitehead has forcibly set down for us, every portion of the universe would seem to be in mutual interaction with every other.

We have no way of picturing such centres of energy, save by an effort of the scientific imagination. They have, however, certain properties which we can measure in the shape of mass, and others which we can get hold of by measuring distances and speeds. Thus, far from the so-called primary qualities being peculiarly *in* matter, we simply take over the idea of mass, distance, and speed from our everyday experience of matter as it presents itself to our senses; we find we can use them to get at and measure certain aspects of the behaviour of these ultimate units of the world; and from the knowledge we thus get of their behaviour, we can build up some doubtless rather inadequate picture of what they really are like. It is only for matter as perceived by our senses—an orange, a table, a stone—that the primary qualities are any more real than the secondary ones. To the physicist, the table is an arrangement of energy, more condensed at innumerable tiny centres, reduced almost to blank space between the centres; and in regard to that picture of the table, the ideas of mass and size are just as much put in by man's mind as is the table's colour, just as much the consequences of the way man works as is the colour the consequence of his eye containing certain kinds of pigment in its cells.

There is another point. The development of relativity theory has shown that these so-called primary qualities are not even unchangeable. For instance, an object going at a very high speed changes slightly both in mass and shape. If from the point of view of the physicist, the properties of matter which we call mass and extension turn out to be merely certain measurable effects of the underlying arrangement of energy-centres, from the point of view of the mathematical philosopher like Einstein, they turn out to be variable properties of a single system—what he calls the space-time continuum—in which matter, space, and time are all inextricably blended, of which they are all merely aspects which we artificially isolate in our thinking. In either case, they disappear as primary essential qualities of matter, and remain only as those manifestations which we human beings can most conveniently get hold of by accurate measurement.

Does all this have any real bearing on practical

life, on us as actual human beings? Is it not all too abstruse and fine-spun?

I do not think so. It seems to me to have quite definite bearings upon various ideas which, consciously or unconsciously, go to make up our general attitude to life; and of course our general attitude to life must in the long run influence the way we live.

To begin with it does in a certain sense put man back in the central position from which science in the beginning had pushed him out. Obviously it is not the identical position. Science gives no support to the idea that man is in any sense a privileged being living in the physical centre of the universe, with the rest of the world created for his use or pleasure. But it establishes human beings as the highest things of which we have knowledge; and it establishes human mind as the one agency which brings order out of the mere chaotic hurly-burly of experience. Either we must give up in despair, or we must trust our own nature. The human mind has created sciences, religions, arts, mathematics, philosophies, moralities. On the whole, there has been definite progress in these constructions of our mind. We therefore have more reason for confidence than for despair. But it must be a confidence in ourselves and our own human powers, not an appeal to something external. It is in that sense that man is re-established in a central position.

And of course this way of thinking brings mind

into the scientific picture. The physicist sees the world as an assemblage of matter and radiation, but forgets to take proper account of the fact that he is able to see it and reason about it at all. In his picture, human mind remains a mere spectator, outside the drama, and wholly unaccounted for. But if, as all physiology tends to show, man is not just a body plus a mind, but a body-mind, body and mind being two aspects of his single nature; and if, as all general biology tends to show, he has evolved from lower forms of life, and life in the final analysis has evolved from matter which was not alive; why, then, something of the same general nature as mind must exist not only in other forms of life but even in lifeless matter.

Thus knowledge and feeling and will are not just something tacked on, so to speak, to a mechanical universe, but the universe is seen to comprise two aspects, one objective and mechanical, the other subjective and concerned with mental and emotional and spiritual happenings: and neither is more real or true than the other.

Finally, mind is not just static. It changes: it, too, like body, is evolving. And the ideas which man's mind hammers out concerning the world he inhabits are in the mental sphere like machines in the physical sphere. They do a particular piece of work more or less adequately. Just as modern hydro-electric plants or textile mills are great improvements on water-wheels or hand-looms, so our mental machinery for obtaining, ordering, and controlling knowledge and thought has improved enormously from what it was among the cave-men or in the time of Moses, in its evolution to the present-day world picture based on science. But equally, like our machines of metal and glass and electricity, it is capable of indefinite further improvement.

The straightforward materialism of science from the seventeenth to the nineteenth century rid us of the idea of magic, and took purpose out of nature. It was no longer necessary or even reasonable to suppose that the stars required guidance in their courses—or at least, any more guidance than a stone dropping to the earth or a stream running downhill. It was no longer necessary or reasonable to imagine that plants and animals, including man, had been specially created, when variation and natural selection would account for their evolution. It became as illogical to pray for rain as it would have to revive the practice, which once seemed wholly natural and sensible, of making sacrifices and carrying out fertility rites to make the crops grow. It destroyed the idea of infallible dogma in the intellectual sphere, and put in its place the conception of a slowly growing, changing body of knowledge.

But there it stopped. It was still in some ways the unchanging spectator outside the world. It could not see itself clearly as part of the evolving universe, it could not yet grasp that its ideas and its very method of thinking about things must change as result of its interaction with new knowledge. Further, it had been so concerned with the intellectual sphere that it had hardly begun to extend its ideas into aesthetics and morals. Most nineteenth-century scientists, for instance, though rejecting the idea of a fixed body of intellectual beliefs, still clung to the idea of a permanently fixed moral code—which happened of course to be that of their own age, nation, and class.

To-day we are beginning to see that the idea of the Absolute, whether in truth or beauty or virtue, is no more and no less than this: it is a necessary consequence of the human faculty (the greatest single difference between man and lower animals) of being able to think in terms of abstract concepts. Once man can say "this is true and that is false," he is setting up, even if often unconsciously, a standard of truth in the abstract. Once he can feel "this is wrong," he has set up an abstract standard of right. But it is an abstraction: the actual truth which he possesses is always only more or less true, the actual morality which he practises only relative to the ideas and the circumstances of his time.

The physicist like Sir James Jeans tells us that mathematical analysis comes nearest to describing reality; and he feels driven to postulate a divine creator or ruler who is responsible for the mathematical order in the universe. But just here the biologist, with his relativist view, feels very suspicious. Surely the physicist, just because he is accustomed to leave mind out of his scientific picture, and yet because mind just refuses to be left out altogether, has put it back again in the form of a divine mind. But the mathematical order in the universe can just as well be thought of as merely the product of the mathematician's wonderful analysis, just as scientific laws are not laws in the ordinary sense, imposed from outside by a law-giver, but are simply the most convenient ways which the scientist can find of describing how things work.

The biologist is tempted to say that Sir James Jeans finds a mathematical divinity ruling the universe just because he himself is such a good mathematician—another example of the human tendency, as Voltaire put it, of creating God in man's image. So Paley, impressed with the evidences of purposeful design in nature (evidences which Darwin later showed could better be explained without conscious purpose, by natural selection), made of the Deity a Divine Artificer; so the early warlike Jews, before the time of the prophets, made of Jehovah a jealous and wrathful divinity.

His own picture is rather a different one. He does not believe that the present state of our knowledge permits any deductions as to the ultimate nature of the universe, its first creation, or final fate, its possible purpose.

Scientific and mathematical laws are one of the ways of our thinking about nature's happenings.

They are the most convenient way of abstracting reality in terms of pure intellect, and also the most convenient way, in the long run, of securing practical control over external nature.

Art is another method of our minds for dealing with phenomena; and religion is yet another. Any one of these ways can be more or less good and true in its own sphere; but however true they may be in their own sphere, they do not and cannot apply to the others in their sphere. And of course, that being so, life is more than science or art or religion alone, and indeed more than a mere addition of them and other separate faculties of life.

In any case, the absoluteness of scientific truth, or religious feeling, or artistic rightness, is something which derives from us. The only immediate reality we know is the stream of raw experience. Science is but one way of arranging this experience in accordance with the laws of our thought. The scientific picture, like any other ordered picture of the universe, is an abstraction.

But the biologist can go one step further. He does not feel able either to assert or to deny that mind, in the shape of a universal or a divine mind, is behind the changing universe he knows. But he can see that mind is an integral part of that universe. Something of the nature of mind must inhere in the essence of things. Under the particular conditions found on this planet the pressure of circumstances has forced mind to become more and more important and elaborate until finally in man it has become self-conscious and the most important single characteristic of the stock.

Life may be a consciously planned experiment on the part of a divine mind—or it may not. But in any case it is legitimate for us to say, on the basis of the known history of life, that mind has become the great progressive feature of life's evolutionary trend. So that, even if our art and religion and science are only our own ways of arranging the jumble of experience, yet in attempting these arrangements we are carrying on with the main trend of evolution. The biologist finds it exceedingly difficult to believe with the pessimists and the sceptics that human life means nothing. It is part of a larger whole, and of a whole with a main upward movement. To continue that trend is to fulfil evolutionary destiny. Clear thinking, deeper feeling, and stronger willing are the chief means of achieving that end. That, I think, is the biologist's chief contribution to our changing picture of the world.

PART III

WHAT IS CIVILIZATION?

BERTRAND RUSSELL

ALDOUS HUXLEY

HUGH I'A. FAUSSET

HILAIRE BELLOC

J. B. S. HALDANE

OLIVER LODGE

BERTRAND RUSSELL

1. THE SCIENTIFIC SOCIETY

THE influence of science upon the everyday life of ordinary men and women is already profound, but is likely to become much more so in the not very distant future. Science is gradually transforming social life in ways which call for new forms of society, and which demand new qualities in eminent citizens. Apart from all detail, the two chief changes that are being brought about by science are the increased importance of experts and the more organic character of human society. Of these two the first, namely, the importance of experts, is more obvious than the other. Nevertheless a few words must be said about it, since it is making effective democracy increasingly difficult. All the everyday apparatus of modern urban life—power stations, electric trains, telephones, electric light, etc.—involves scientific knowledge possessed by only a small minority of the population. By killing off a suitably selected 1 per cent of any modern nation, its present mode of life could be made impossible. In matters more directly concerned with government, the same thing is true in an even higher degree. The art of war, although it still requires soldiers, depends much more upon the scientific inventor than upon the man who risks his life in the face of the enemy.

The art of banking, as we have lately been learning to our cost, is so intricate that even the recognized experts make a mess of it. For all these reasons democracy is less effective in a highly developed industrial society than it can be in a simpler agricultural community, since there are many questions which ordinary men and women cannot understand, and in regard to which they are compelled willy-nilly to accept the opinions of specialists. The importance of experts is likely to increase rather than diminish as the part played by science in daily life grows greater. We must therefore expect that, in the future, government by experts will largely replace government by the will of the people, even if the outward forms of democracy are preserved intact.

Even more important is the increasingly organic character of society. When I say that society is increasingly organic I mean that the acts of one individual or set of individuals tend more and more to have effects upon other individuals, perhaps in distant parts of the world. The doings of a primitive peasant who lives upon his own produce are of little importance except to himself and his family. But where modern industrial methods prevail, the world has become a single economic unit. The political passions of the Chinese may cause destitution in Lancashire; the prejudices of the American Middle West determine the character of the cinemas offered for the amusement of Europe; the existence of oil in the Middle East profoundly affects international politics, and at one time even caused friction between America and Great Britain.

In proportion as society becomes more organic, it is necessary that it should be more organized. A society is organic in proportion as what happens to one part has effects upon another part; it is organized when these effects are determined by relation to the welfare of the whole. The human body, for example, is at all times organic, and in health is also organized. But in certain kinds of disease (for example, cancer), while it remains organic it ceases to be organized. Modern society is very much more organic than the society of two hundred years ago, and is somewhat more organized. But the extent to which it is organized has not increased nearly as fast as the extent to which it is organic. If our scientific civilization is to be stable, it is imperative that it should become much more organized than it is at present; there must be much more deliberate planning and much less left to the haphazard operation of individual impulse. This applies to all kinds of matters: municipal, national, and international. London, historically considered, is a collection of villages which have gradually coalesced, but in the present day it is a unit, and one of its most important features is the means of transport by which men pass from the circumference to the centre in the morning, and from the centre to the circumference in the evening. If London had been deliberately planned to suit its present needs, the streets would all be straight and all thoroughfares.

The most important respects in which organization is at present deficient are the economic anarchy due to undirected enterprise, and the political anarchy due to unrestricted national sovereignty. The various parts of the world have become economically interdependent, but there is no international economic organization either of production or of banking. Each nation wishes to produce everything itself, with the result that the industrial plant in the world is capable of producing much more than the world is able to consume. The increased productivity of labour resulting from modern technique has therefore resulted in bankruptcy for employers and unemployment for wage-earners, when, if there had been any international organization of production, it might have resulted in wealth for employers, and full wages, with shorter hours, for wage-earners.

The economic anarchy in the world has proved disastrous in recent years, but the political anarchy consisting in the absence of an international government is likely to prove even more disastrous. Immense scientific skill has been applied to the technique of war, and unfortunately in recent years methods of attack have made much greater progress than methods of defence. The next war is therefore likely to prove far more destructive than what, as yet, we still call the Great War. Scientific civilization cannot survive unless large-scale wars can be prevented, and large-scale wars cannot be prevented except by the establishment of a world government possessing the only effective armed force in the world. The indications at present are that men would rather see civilization perish than adopt this means of preserving it. But it is probable that after the next war they will change their minds; if they do not, scientific civilization will disappear.

Owing to the increasing need of organization, a scientific society, if it is to be stable, will necessarily involve a diminution of individual liberty as compared with the societies of the past. This is regrettable, but apparently unavoidable. There will be, however, such important compensations that, on the balance, we may expect an increase in human happiness. Science has already done a very great deal to lengthen human life and to diminish disease. In this respect it is sure to do more in the near future. It has not yet destroyed poverty and the fear of destitution, but it has created the technical possibility of achieving this result. In the pre-scientific ages the total produce of human labour yielded so little above a bare subsistence that only a very small minority could enjoy tolerable comfort. Nowadays the productivity of labour is such that, given a wise international organization of the world's productive efforts, it would be possible within a generation to secure tolerable comfort for everyone without very long hours of labour. This possibility we owe to science. The fact that it is not realized we owe to stupidity and inertia. If men acquire the wisdom to utilize existing knowledge to the full, they may, within the next hundred years, establish throughout the world a community wholly freed from the dangers of war and poverty, and at the same time healthier and longer-lived than even the best communities now existing. This cannot, however, be achieved without a considerable surrender of liberty, both in the economic and the political sphere, since it requires an international government and international control over production and distribution.

No civilization worthy of the name can be *merely* scientific. Scientific technique is concerned with the mechanism of life: it can prevent evils, but cannot create positive goods. It can diminish illness, but cannot tell a man what he shall do with health; it can cure poverty, but cannot tell a man how he shall spend wealth; it can prevent war, but cannot tell a man what form of adventure or heroism he is to put in its place. Science considered as the pursuit of knowledge is something different from scientific technique, and deserves a high place among the ends of life, but among these it is only one of several. At least equal to it are the creation and enjoyment of beauty, the joy of life and human affection. A scientific society which did not promote these things could not be considered positively excellent, even if it were to eliminate much of the pain and misery from which mankind has hitherto suffered. The society of the future, assuming that we can escape a cataclysm, will be more and more a thing deliberately planned rather than a spontaneous natural growth. But in this deliberate planning the technician, if he is to be wise, will have to take account of human values which lie outside the immediate scope of his technique. Men of the administrative type will necessarily have more power in the scientific society than they have had at any previous stage of the world's history, and if they are not to abuse this power, their education will have to be carefully directed towards giving them a breadth of outlook which at present is not always to be found among such men. There is in this a danger, but it is a danger which can be avoided if men are sufficiently aware of it.

The scientific society differs from the unscientific society fundamentally through the fact that in the scientific society men know better how to realize their desires. If this is to be a boon to mankind, it is necessary that men's desires should be constructive rather than destructive. To secure that this shall be the case is itself a problem for science, namely for the sciences of psychology and education, to which perhaps I should add physiology. Power in the hands of men whose passions are anarchic is dangerous, and I fear that mankind will have to pass through some very painful ordeals before they learn to use scientific power wisely. But I cannot doubt that in the end the lesson will be learnt, and that when it has been learnt, the human race will find itself emancipated from many of the greatest evils that have afflicted it in all past ages. The immediate outlook is uncertain and full of danger, but the more distant outlook permits hopes which would have seemed fantastic in any earlier time.

2. ECONOMISTS, SCIENTISTS, AND HUMANISTS

THERE are certain values which we feel to be absolute. Truth is one of them. We have an immediate conviction of its high, its supreme importance. Science is the organized search for truth and, as such, must be looked upon as an end in itself, requiring no further justification than its own existence. But truth about the nature of things gives us, when discovered, a certain control over those things. Science is power as well as truth. Besides being an end in itself, it is a means to other ends. Science as an end in itself directly concerns only scientific workers and philosophers. As a means, it concerns every member of a civilized community. I propose to discuss here science as a means to ulterior ends—ends which may be summed up in the single vague and comprehensive word, "Civilization."

Our civilization, as each one of us is uncomfortably aware, is passing through a time of crisis. Why should this be? What are the causes of our present troubles? They are all due, in the last resort, to the fact that science has been applied to human affairs, but not applied adequately or consistently.

In the past, man's worst enemy was Nature. He lived under the continual threat of famine and pestilence; a wet summer could bring death to whole nations, and every winter was a menace. Mountains stood like a barrier between people and people; a sea was less a highway than an impassable division. To-day, Nature, though still an enemy, is an enemy almost completely conquered. Modern agriculture assures us of an ample food supply. Modern transportation has made the resources of the entire planet accessible to all its inhabitants. Modern medicine and sanitation allow dense populations to cover the ground without risk of pestilence. True, we are still at the mercy of the more violent natural convulsions. Against earthquake, flood, and hurricane man has, as yet, devised no adequate protection. But these major cataclysms are rare. At most times, Nature is no longer formidable; she has been subdued.

Our present troubles are not due to Nature. They are entirely artificial, genuinely home-made. The very arts and sciences which we have used to conquer Nature have turned on their creators and are now conquering us. The present crisis is of our own making; we have brought it on ourselves by allowing our mechanical and agricultural science to develop more rapidly than our economic science. We cannot buy what we produce and are therefore compelled to keep our factories idle and let our fields lie fallow. Millions are hungry, but wheat has to be thrown into the sea. This is where, at the moment, science has brought us.

What is the remedy? Tolstoyans and Gandhi-ites tell us that we must "return to Nature"—in other words, abandon science altogether and live like primitives or, at best, in the style of our medieval ancestors. The trouble with this advice is that it cannot be followed—or rather that it can only be followed if we are prepared to sacrifice at least eight or nine hundred million human lives. Science, in the form of modern industrial and agricultural technique, has allowed the world's population to double itself in about three generations. If we abolish science and "return to Nature," the population will revert to what it was—and revert, not in a hundred years, but in as many weeks. Famine and pestilence do their work with exemplary celerity. Tolstoy and Gandhi are professed humanitarians; but they advocate a slaughter, compared with which the massacres of Tinur and Jinghiz Khan seem almost imperceptibly trivial.

No, back to Nature is not practical politics. The only cure for science is more science, not less. We are suffering from the effects of a little science badly applied. The remedy is a lot of science, well applied.

Everyone admits in principle that human activities must be regulated scientifically; but when it comes to applying this principle, two questions arise. Science, in the present context, is a means to an end: but what end? That is the first question. And (this is the second question) by whom is this instrument to be used? Who is to wield the power which science gives?

To define the ideal human society is not too difficult. It is a society whose constituent members are all physically, intellectually, and morally of the best quality; a society so organized that no individual shall be unjustly treated or compelled to waste or bury his talents; a society which gives its members the greatest possible amount of individual liberty, but at the same time provides them with the most satisfying incentives to altruistic effort; a society not static, but deliberately progressive, consciously tending towards the realization of the highest human aspirations. Science might be made a means for the creation of such a society—but only on condition that the powers it confers be used by rulers inspired by what I may call humanistic ideals.

Our present crisis is mainly and most obviously economic. The fact is dangerous; for it means that the ends pursued by our rulers, at any rate in the immediate future, will be primarily economic ends. It means that the instrument of science will be used by men primarily interested in economics and only secondarily, if at all, in the higher humanistic values.

I have described the humanist's earthly paradise. What is the economist's ideal society? Briefly it is one where there is the maximum of stability and uniformity. The economist wants stability because, once you set machinery going, it is hopelessly uneconomic to let it stop or run irregularly. Also industrialists and financiers must be able to look forward with confidence; in a stable world the machine is able to go on running steadily. Again, the economist wants uniformity, because the most profitable form of mechanical production is mass-production. The mass-producer's first need is a wide market—which means, in other words, the greatest possible number of people with the fewest possible number of tastes and needs.

Now stability and a certain amount of uniformity are essential pre-requisites to any rational plan for improving the quality of civilization. They are means to ends, not ends in themselves. But it is precisely as ends in themselves that the economist-rulers of the immediate future are likely to conceive them. It is easy to imagine an oligarchy of industrialists and financiers using all the resources of science first to secure world-wide stability and uniformity and then, in the interests of production, to keep the world stable and uniform. The aim of the economist will be to make the world safe for political economy— to train up a race, not of perfect human beings, but of perfect mass-producers and mass-consumers. One of the things economist-rulers would be almost bound to do is to suppress science itself. Once stability has been attained, further scientific research could not be allowed. For nothing is more subversive than knowledge. So long as scientific research goes on, society stands poised above a potential succession of earthquakes. Any day some new discovery may make all existing equipment obsolete, may revolutionize all existing technique, or else, by changing man's physiological habits, radically alter his whole way of thinking and feeling. Having first made use of science, economist-rulers would find themselves forced to destroy it. Even humanist rulers might often have to forbid the application of certain discoveries. Let us suppose, for example, that a method has been discovered for producing all food synthetically. Humanist-rulers might feel justified in forbidding the application of the discovery on the grounds that agricultural life was humanistically valuable.

But these are remote speculations. Let us try to guess how the resources of science might be used or abused by different types of rulers in the nearer future.

I will begin with psychology, the science which concerns us more closely and intimately than any other—the science whose subject-matter is the human mind itself. In a rather crude and ineffective way psychological knowledge is already applied to the problems of government. It was shown during the war that propaganda—which is the art of influencing the mind—could become one of the major instruments of national policy. Profiting by war-time experience, the rulers of Russia and Fascist Italy are systematically using this psychological weapon to create new types of civilization. Even in conservative England our rulers have not disdained to take a leaf out of the Soviet and Fascist book. Systematic mass-suggestion by wireless and poster played a very important part, as we all know, in the last election and during the Buy-British campaign of the Empire Marketing Board.

Propaganda is still relatively inefficient even in countries like Italy and Russia, where the state controls all the existing instruments of mass-suggestion, from education to the movies and the Press. But psychological science teaches how it could be made almost irresistibly effective. Freud and his followers have shown how profoundly important to us are the events of the first few months and years of our existence; have proved that our adult mentality, our whole way of thinking and feeling, our entire philosophy of life may be shaped and moulded by what we experience in earliest childhood. Following another line of research, the great Russian biologist, Pavlov, and the American Behaviourists have shown how easy it is, with animals and very young children, to form conditioned reflexes which habit soon hardens into what we are loosely accustomed to call "instinctive" patterns of behaviour. Such are the scientific facts waiting to be applied to the solution of political problems. Rulers have only to devise some scheme for laying their hands on new-born babies to be able to impose on their people almost any behaviour pattern they like. No serious practical difficulties stand in the way of such a plan. One of these days some apparently beneficent and humanitarian government will create a comprehensive system of state crèches and baby farms; and—with a little systematic conditioning of infant reflexes—it will have the fate of its future subjects in its hands. From the baby farm the already thoroughly contioned infant will pass to the state school. He will grow up reading state newspapers, listening to state wireless, looking at state cinemas and theatres. By the time he reaches what is somewhat ironically called the age of reason, he will be wholly unable to think for himself. None but approved state ideas will ever even occur to him. This will make the overt use of force quite unnecessary. Dictatorship, as a form of government, may have, in the immediate future, a brief spell of popularity. In times of crisis like the present, strong government is probably necessary. But once the position has been stabilized and, above all, once our rulers have been educated up to the point of realizing the extent of the power which psychological science has placed in their hands, strong government will cease to be necessary. When every member of the community has been conditioned from earliest childhood to think as his rulers desire him to think, dictatorship can be abandoned. The rulers will re-establish democratic forms, quite confident that the sovereign people will always vote as they themselves intend it to vote. And the sovereign people will go to the polling booths firmly believing itself to be exercising a free and rational choice, but in fact absolutely predestined by a life-long course of scientifically designed propaganda. Its choice will be made by an inward, psychological compulsion much more powerful than any pressure of physical force from without.

For the economist-ruler, scientific propaganda will seem a heaven-sent instrument. He will use it to train up that race of perfect producers and consumers of which industry has need. He will find it invaluable for producing and preserving that stability and uniformity, without which machines cannot be used to their maximum advantage. By means of it a creed will be inculcated, racial and individual idiosyncrasies as far as possible smoothed out, contentment and conformity incessantly preached. Indeed, scientific propaganda may enable future rulers to do what the medieval popes and emperors tried but failed to achieve. They may actually succeed in creating a great world-wide community united by common beliefs and aspirations, common wants, tastes and thoughts. It will be a Holy Roman Empire minus the holiness, a Christendom but without the Christianity ... or if nominally Christian, Christian in a way that neither the primitive convert, nor the medieval Catholic, nor the later Protestant would recognize as Christian.

What will be the attitude of the humanist towards scientific propaganda? Fundamentally, I think, he would be opposed to it. For if it were thoroughly scientific and efficient, scientific propaganda would obviously be quite incompatible with personal liberty. Now personal liberty is, for the humanist, something of the highest value. He believes that, on the whole, it is better to go wrong in freedom than go right in chains—even if the chains are imponderable, even if they are not felt by the prisoner to be chains. Nevertheless, it may be that circumstances will compel the humanist to resort to scientific propaganda, just as they may compel the liberal to resort to dictatorship. Any form of order is better than chaos. Our civilization is menaced with total collapse. Dictatorship and scientific propaganda may provide the only means for saving humanity from the miseries of anarchy. The liberal and the humanist may have to choose the lesser of two evils and, sacrificing liberty, at any rate for a time, choose dictatorship and scientific propaganda as an alternative to collapse. Again, the humanist will have to remember that propaganda is a substitute for force in general and war in particular. It would certainly be worth forgoing a great deal of liberty for the sake of peace.

I have dwelt at some length on propaganda because it seems to me that, without it, there can be no large-scale application of scientific knowledge to human affairs. Psychology is the key science. Many of the possible applications of biology, for example, are so startling that they must be prepared for by a regular barrage of propaganda. Sprung too suddenly on the world, they would be passionately resisted. Let us now consider a few of these possible applications of science, speculating as before how they might be used by the humanist or abused by the economist.

Biologists have collected a very considerable amount of information on the subject of heredity and are steadily adding to their store. So far as our knowledge goes, negative eugenics—the sterilization of the unfit—might already be practised with tolerable safety. On the positive side we are still very ignorant—though we know enough, thanks to R. A. Fisher's admirable work, to foresee the rapid deterioration, unless we take remedial measures, of the whole West European stock. Eugenics are not yet practical politics. But propaganda could easily make them practical politics, while increase of knowledge will make them also purposive and far-sighted politics.

The humanist would see in eugenics an instrument for giving to an ever-widening circle of men and women those heritable qualities of mind and body which are, by his highest standards, the most desirable. But what of the economist-ruler? Would he necessarily be anxious to improve the race? By no means necessarily. He might actually wish to deteriorate it. His ideal, we must remember, is not the perfect human being, but the perfect mass-producer and mass-consumer. Now perfect human beings probably make very bad mass-producers. It is quite on the cards that industrialists will find, as machinery is made more fool-proof, that the great majority of jobs can be better performed by stupid people than by intelligent ones. Again, stupid people are probably the state's least troublesome subjects, and a society composed in the main of stupid people is more likely to be stable than one with a high proportion of intelligent people. The economist-ruler would therefore be tempted to use the knowledge of genetics, not for eugenic, but for dysgenic purposes —for the deliberate lowering of the average mental standard. True this would have to be accompanied by the special breeding and training of a small caste of experts, without whom a scientific civilization cannot exist. Here, incidentally, I may remark that in a scientific civilization society must be organized on a caste basis. The rulers and their advisory experts will be a kind of Brahmins controlling, in virtue of a special and mysterious knowledge, vast hordes of the intellectual equivalents of Sudras and Untouchables.

What is true of applied genetics is true, *mutatis mutandis*, of applied bio-chemistry and pharmacology. Our knowledge of what can be done by means of drugs is still rudimentary. It may be possible, for example, to modify profoundly men's character, temperament, and intelligence by administering suitable chemicals at suitable moments. Yet once more, the same knowledge will be used by the humanist and the economist in profoundly different ways.

I will not discuss the possible effects on human beings of other scientific discoveries. History shows that almost any new acquisition of knowledge may be made the basis of important practical applications. The abstruse researches of Faraday and Clerk Maxwell have resulted, among other things, in the jazz band at the Savoy Hotel being audible at Timbuctoo. Not a very probable result, you must admit. But then the course of events takes no account of verisimilitude. Fiction has to be probable; fact does not.

And here I should like to make what to me seems an important point. We are unable to foresee what discoveries in pure science will be applied to human life. But equally we are unable to foresee all the results of any given application of science. Certain particular ends may be envisaged by the man who applies scientific knowledge, and these ends may, in fact, be attained. But almost inevitably other ends, not foreseen, will have been attained at the same time. For example, when Bradlaugh and Mrs. Besant broadcast the medical knowledge which has been applied as birth-control, their intention was that families should be reduced in size. Their action produced its intended effect; but it also produced effects which they certainly did not intend. For example, it forced architects to build tall blocks of five-roomed flats, rather than long rows of fifteen-roomed houses; and it compelled farmers to breed small cattle rather than large ones. A century ago prize bulls weighed as much as two tons; to-day small families require small joints of meat and prize bulls weigh about half a ton. These unintended effects of birth-control are not particularly important or significant. But it often happens that the unintended effects of an action are much more considerable than the intended ones. The application of science to human life has already produced a large crop of unintended effects, some of which are highly undesirable. Science increases our powers of foretelling the future; but we may be quite sure that it will be a very long time before the unintended effect will be altogether eliminated. Nor must we forget that these unintended effects will follow actions undertaken with the highest possible motives. The well-meaning humanist is as likely to give people an unpleasant surprise as the ill-meaning economist. Against unpleasant surprises there is no remedy. Each unexpected situation must be dealt with individually, as it turns up. We can only hope that it will not prove too unpleasant, and that increasing knowledge will permit of more accurate foresight.

I will only add this by way of summary and epilogue. Science in itself is morally neutral; it comes good or evil according as it is applied. Ideally, science should be applied by humanists. In this case it would be good. In actual fact it is more likely to be applied by economists, and so to turn out, if not wholly bad, at any rate a very mixed blessing. It rests with us and our descendants to decide whether we shall use the unprecedented power which science gives us for good or for bad purposes. It is in our hands to choose wisely or unwisely. Alas, that wisdom should be so much harder to come by than knowledge!

3. SCIENCE AND THE SELF

In the first contribution to this series Bertrand Russell viewed with some complacence the scientific society of the future. He admitted that we have not yet organized the world as a whole, but he was hopeful that through science we should eventually do so. I am not concerned, however, at the moment with the question of world organization, important as that is, but with the inner life of the individual. Bertrand Russell, indeed, reminded us that the things which give positive excellence to human life are in the mind and heart, not in the outward mechanism. And it is upon the mind and heart and the relation of science to them that I wish to concentrate.

The questions I am going to ask and try to answer are these: Has natural science, despite all its mental and material conquests, impoverished our real life? And, if so, must it continue to impoverish it? Is its method of acquiring knowledge the only true method? Or is it fatally partial and one-sided?

This last may seem a surprising question, for during the last hundred years the scientist has popularized the view that he alone was exercising his reason aright, and that those who claimed to arrive at knowledge by other methods than his were in different degrees clinging to false illusions because they were too weak to face the truth.

He is perhaps less certain of this to-day. Nevertheless the assumption is still prevalent. Yet however intelligible such an attitude was in mid-Victorian times, when Natural Science was fighting a necessary battle against religious obscurantism, it represents itself to-day an equally dangerous kind of dogmatism.

For the scientist's method of acquiring knowledge is not the only valid one. His aim is to reduce the human mind to a sensitive machine which sorts the facts given to it by observation, measures them in relation to one another, and arranges them in a correct pattern according to its own inherent logic. When new facts are discovered, the pattern is modified to include them. But it is always the simplest pattern into which the facts will fit.

Obviously in such a process the mind can never be a mere machine. An act of will is involved on the part of the searcher and even a sense of form, akin to that of the artist. But this personal element has been generally denied by the scientist. He has insistently claimed that his approach to truth is purely impersonal. He has striven in his researches to exclude every desire or interest of his own, and the better to ensure this he has taken elaborate external precautions against personal prejudice.

A necessary result of this attempt to acquire exact knowledge independent of any personal act of knowing is that the scientist, as Professor Levy has already reminded us, can only deal with what is constant and common to all observers. He is compelled to disregard the unique reality of an object and reduce it to a mere instance in a series of instances. All qualitative values disappear beneath a ruthless classification and all living form perishes in abstract formulas.

This process of abstraction is displayed perhaps most notably to-day in the subtle but tenuous formulas of the mathematical physicists. But while one may well admire the way in which the material world has dissolved beneath their measuring-rods, their attempt to produce a purely intellectual representation of the universe has inevitably resulted in what is at best only a ghostly skeleton of reality. And every one of the physical sciences which attempts an exclusively intellectual approach to Nature suffers under the same disability. Since Ultimate Reality cannot be calculated, since it must be immediately experienced, they can never really know life, but only something of the mechanism of life's expression. But if ultimate Reality has escaped and must always escape the physical scientist, he can justly claim that his method has proved remarkably successful in its own relative domain. And although the knowledge which he has thus acquired has not noticeably increased human happiness, it has made it possible for man to master to some extent his physical conditions, to alleviate physical pain, and to exploit for his own material benefit the forces of Nature.

But there is another and older theory of knowledge. According to it, we cannot know the reality of anything unless we enter into it imaginatively, unless we wholly identify ourselves with it and realize it from within. To achieve real knowledge, therefore, it is necessary not merely to observe and co-ordinate facts, but to live the truth. Knowledge, in short, depends upon the quality of being possessed by him who seeks to know. To know better, it is necessary to become different. For the more deeply harmonized are a man's faculties of feeling and thought, the finer and more fundamental are his powers of achieving contact with reality.

This is the science of the poet, the mystic, and the seer, and of all who try to know life with their whole being. And we have an elementary example of such integrity of being in the simple, necessary response to life of the child and the peasant.

I am not, however, suggesting that the progressive claims of modern science can be disproved by pointing to the humble virtues of the peasant or the child. For the rational self-consciousness from which the world is suffering is necessary to human development. And all attempts to revert to childhood are inevitably doomed to failure.

Nevertheless simple people do offer us a suggestive example, on the instinctive level, of wholeness. For however undeveloped their powers of conscious intelligence may be, the knowledge which they possess and the thought which they exercise are grounded in their very being. And this is true, on a more advanced level, of all creative thought. Unlike the critical analysis of science, it is an expression of the whole being.

Such spiritual perception or imaginative knowledge is little regarded in the West to-day, because we have been witnessing during the last hundred years the culmination of a process clearly traceable from the Renaissance in Europe and the Reformation in England. It was as inevitable a process as that which occurs in every individual who in passing from childhood to youth is inwardly divided.

Out of this division a richer and deeper unity may be ultimately achieved. But meanwhile, because the individual is at conflict with himself, he is at cross-purposes with life. He is either stricken with indecision or he asserts his personal will against life, denying it as a whole in the interests of one of its parts. Consequently his soul loses all contact with its depths and he becomes mentally expert but superficial.

And Western civilization for the last hundred years clearly reflects such a state, a state in which the personal will of the individual has lacked any creative centre, so that he has sought increasingly his own private gain or glorification. And because modern man has become thus uncentred, modern civilization has been full of discord and aimlessness. With immense resources of wealth and power, it has lacked unity of design or purpose.

And the basic weakness of natural science, so far as it claims to cure the disease of civilization, is that it suffers from the disease itself. In its exploitation of one faculty, the intellect, at the expense of all the others, it has reflected and aggravated the separation of knowledge from being. It has, indeed, affirmed the unity of physical Nature, but it has denied that higher unity to which man as a creative spirit belongs. Certainly, as Aldous Huxley said, science is morally neutral. But it is also spiritually blind. Concerned itself only with the processes reflected in physical matter, it has assumed and popularized the view that these are alone real. Because it can only deal with physical organisms, it has tended to reduce man to the same physical level as frogs or rabbits. And so far as modern man has accepted the scientific view that his body is more real than his soul, he has become the slave of external things and secondary conditions instead of realizing the inward freedom that comes of obedience to the commands of the higher self.

Our greatest need to-day, therefore, is not to deny the intellect, but to make it more profound. And we can only do this by recognizing that it must be subordinated to something more complete and essential than itself.

The problem of knowledge and of life, and so of civilization, is, in fact, ultimately, as all the great mystics and spiritual teachers have insisted, a moral one. They admit that man at a certain stage of his development falls into sin or division. But they affirm out of their own experience that by sustained effort and self-culture, by humbling himself to life and at the same time exercising and perfecting all his faculties, man can bring his consciousness again into a state of unity. And that in such a state of unity not only, to use Blake's words, are the doors of perception cleansed and the eternal significance of every particular divined, but at the same time a true disinterestedness is achieved.

The claim of the scientist to be disinterested beyond all other men has been, however, so frequently advanced and generally accepted that it may be well to consider it for a moment. In a famous letter to Charles Kingsley, Thomas Henry Huxley wrote: "Science seems to me to teach in the highest and strongest manner the great truth which is embodied in the Christian conception of entire surrender to the will of God. Sit down before fact as a little child, be prepared to give up every preconceived notion, follow humbly wherever and to whatever abysses Nature leads, or you shall learn nothing."

It is hardly necessary to say that the manner in which a little child sits down before fact is very different from that of even the most conscientious natural scientist. For the child's relation to fact is not one of mental observance, but of such sensitive and whole-hearted absorption that it is not perhaps too much to say that no facts exist for him.

And the same distinction may be drawn between the mystic's or the artist's surrender of his whole self to the creative will and the precautions which the scientist takes against personal bias. The one is a moral and entire, the other only a mental and partial act. Admittedly this distinction does not apply to the greatest scientists, who have not only obeyed the rules of research, but always possessed, too, the gifts of divination. They, like the artist, by submitting to a technical discipline, have prepared themselves for the creative moment when a truth is given to them. And it was given to them because instead of priding themselves upon their intellectual respect for fact they humbled themselves to reality. But such scientists are as rare as men of disciplined and disinterested vision always are. Moreover they attain to truth not primarily through conforming to a creed and a practice peculiar to science, but because they are truly selfless and so inspired by the creative spirit.

And such selflessness is not achieved by merely surrendering the self to facts and to instruments devised to measure facts.

Outside, indeed, the province of the laboratory, in which impartiality can be technically guaranteed— in Biology, for example, as distinct from Chemistry and Physics, or in such border sciences as Anthropology, Psychology, and Sociology, which deal with life where it has ascended from the purely physical to the human plane—we constantly find that the scientist has projected his personal prejudice into his interpretation of phenomena while claiming to be wholly disinterested.

The blindness of nineteenth-century evolutionists, for instance, to the co-operative principle in Nature was due to an innate combativeness in themselves. Their concentration upon natural selection and the survival of the fittest to the exclusion of creative variation and mutual adaptability reflected their own individual limitations. And in the same way the anthropologists of yesterday explained the life and customs of savages in terms of their own self-assertive consciousness, attributing to primitive man the "tiger qualities" of a predatory civilization. For, as Amiel wrote, "a man only understands what is akin to something already existing in himself."

I would ask you to remember, therefore, in reading Professor Levy's account of scientific method, that the scientist's professed respect for facts, his interpretation and even his recognition of such facts as cannot be measured and tested by retorts and balances, must depend upon the degree of his own real integrity. And scientific method does not enforce such integrity. For the discipline of the laboratory involves no real change of being and no deep culture of the self. It may and does encourage a specialized habit of cautiousness and accuracy. But there is no necessary relation between the sensitiveness of the scientist's instruments and the real sensitiveness of the man himself. Moreover, by delegating sensitiveness to instruments, or exercising it only in a narrow and abstract field, he tends, as even Darwin regretfully admitted, to lose it himself.

It is possible therefore to be a brilliant scientist and yet in feeling to be quite uncivilized.

For what is it to be truly civilized? It is surely to draw upon deep inward resources and at the same time to be finely responsive to one's environment; it is, out of the fullness of a true self, to respect the uniqueness of every living person or creature and to be incapable of exploiting them. It is to co-operate with the spirit of Nature rather than to master her physical processes by intellectual cunning or for selfish ends. It is to live unattached to material things and desires, yet accepting both as means for the expression of a spiritual harmony.

A scientist may well be civilized in this true sense, but only very partially as a scientist. For professionally he is compelled to accept the physical and quantitative aspect of the world as alone real and to deny through all his working hours the validity of anything but a mental and mathematical approach to it.

Modern science, in short, by its insistence that perception should be as much dehumanized as possible and by its consequent blindness to those living realities which escape its measuring instruments, has tended more and more to empty life of real meaning. Doubtless the material world has dissolved before it into a fine-spun web of abstract formulas, but the practical effect of this triumph of the technical mind over matter has been to subordinate man to the machine. For a mechanism is as necessary an offspring of science as an organism is of creative imagination. And although the elaborate and standardized mechanism which science has constructed in the modern world is proof enough of its astonishing mental ingenuity, the individual soul has been increasingly crushed and stifled beneath it. Admittedly the negative side of the modern scientific movement does not affect the disinterested virtue of science at its purest and best. But the effects of such pure science on modern civilization have so far been slight compared to the overwhelming pressure of applied science, and of its offspring, the machine.

Certainly the machine, apart from its productive and labour-saving uses to society, does impose a discipline upon those who serve it. It claims exactness, efficiency, and a subordination of self to its technical demands. But because its technical demands, unlike those of any art or craft, are mechanical, because it denies its servants the right to disciplined self-expression, it tends to reduce them to ciphers, to mere cogs in its ruthlessly rotating wheels. They are cut off from the deep rhythm of life and condemned to a sterile service.

And the evidence of this sterility extends far beyond those who are actually tied through all their working hours to machines. We live in a day when the unique is everywhere being submerged in the uniform, and although we may pride ourselves upon a certain intellectual candour and dexterity and to some extent a concern for physical well-being, these virtues are conditioned and counterbalanced by the fact that we have little desire to raise ourselves to a higher pitch or, indeed, to conform to anything but average standards.

Yet beneath our physical and intellectual activities the deepest needs of the soul remain unsatisfied. And physical science in itself can do nothing to supply them. It can only strive to distract our attention from them. And it does this in a typical way.

For how often are we asked to bow down before the wonders of modern mechanical invention. And at first we cannot but be impressed; nor would I deny that some of these products, which we owe directly to the scientific mind, may be of value as well as utility. But if we consider the matter more carefully, do we not find that the great majority of them supply no really felt need? That by multiplying trivial objects, we have multiplied trivial desires, and that we could live a life more rich in meaning without them? For the deepest need in man or woman is to express the self in some sort of creative activity, however humble. And although the burdened housewife may well be grateful to modern science for certain labour-saving devices, and congested city life would be impossible without scientific organization, is not the emphasis which science lays upon labour-saving and its eagerness to substitute the machine for the person or reduce the person to the machine a denial of true life values?

If science were merely striving to make more tolerable soul-killing conditions which it has helped to create, I would be more willing to approve its endeavours and acclaim its successes. But if we are to believe one of its latest exponents, it is even working to deprive woman of one of the few kinds of labour left in the modern world which can afford a deep creative satisfaction by manufacturing human life in a laboratory. The dangers that attend a self-sufficient quest of knowledge could hardly be better illustrated. Yet such a suggestion is wholly in accordance with the logic of science. To manufacture human life is just as reasonable and right to it as to manufacture poison gas.

But it may be said that the theory of knowledge and of life, which emphasizes the eternal value of the self and the need of wholeness in man's response to experience, has defects of its own when put to the test of practice no less obvious than those of physical science. And this cannot be denied. Modern psychologists have done a real service in showing how often those who claim to be inspired servants of the truth and insist most forcibly upon the absoluteness of their vision are in fact indulging their egotism or compensating themselves for some inner weakness or unsatisfied desire.

Critical detachment is, indeed, essential to truth

if self-delusion is to be avoided. But it should be realized as a means, not an end, as an element in creative activity. In modern science it has been cultivated exclusively and with such energy and pertinacity that it has almost destroyed the creative unity of life. The part has usurped the place of the whole; analysis has eaten not only into the body, but into the spiritual nerve-centres of life.

Yet we have no right to complain of such specializing in itself. Natural science has, indeed, brought light into dark places and led to discoveries which can on the material plane assist us very considerably in building up conditions favourable to creative living, and may even help us to perfect, notably through psychological understanding, that true science of being to which it is itself indifferent.

But it can only do so if it is recognized as a servant, not a master. "Scientific method," in the words of Jung, the most imaginative of modern psychologists, "must serve; it errs when it usurps a throne." For while a true spiritual science can include the mental province of physical science within it, physical science, however rarefied the fruits of its analysis may become, can never in itself regain the unity of perception from which it has broken away. Yet the fact that we have extended and refined our physical knowledge to such a degree may well result in a corresponding richness of consciousness and capacity if we can recover a spiritual unity.

And this can only be achieved by cultivating an

inward life to counteract the immense pressure of material life upon us.

Far from helping us to develop that inward being, to liberate and vindicate that essential self from which true action and true knowledge spring, natural science has inevitably hitherto denied it. And so its mental victories have been won at the cost of our moral defeat. And we shall remain demoralized until we realize that we must creatively unify ourselves before we can create value and unity in the external world.

For no true civilization can be achieved by working adroitly on the surface. It is ultimately conditioned by the spiritual strivings of countless individuals. It has a soul, so far as they have souls. The partial knowledge of natural science has doubtless proved an excellent antidote to timid credulity and an effective weapon against selfish ignorance. But it has been and is in many ways fatal to the growth of the finer spiritual qualities. When, however, it is recognized to be partial, it may prove of real service to men in the task of making themselves and their world a whole.

Both Bertrand Russell and Aldous Huxley have dealt largely with the possibility of curing the disease of civilization from without. Science for them is the doctor, who, as he becomes more skilled in his profession, may eventually cure the patient, if he does not first kill him.

I have tried to suggest that the patient must cure himself, that he can only recover real health by inward effort and obedience to the deepest laws of his nature, and that although he may in his sickness need and profit by the specialist's treatment, the increasing tendency to depend on external aids, the ulterior effects of which are hidden even from the specialist himself, is a very dangerous one. For true life and health can only be realized from within. And although a scientific society might be efficiently constructed by technical experts, it would be a society in which the individual was not more but less himself than in a tribal community. A civilized society can only be created by men and women who are experts, not in some particular science, but in the art and understanding of Life.

4. MAN AND THE MACHINE

MAN seeks truth. He attempts to arrive at reality. He is the only animal that feels this curiosity and acts on it; just as he is also the only animal that laughs, that worships, that speaks and thinks rationally. In other words, he is the only animal that is not an animal.

There are many ways by which man arrives at a truth. He arrives at a moral truth by the conscience, he arrives at a mathematical truth by deduction, he arrives at the truth on beauty and on order by his aesthetic judgment. But one special way, applying only to one sort of truth, is by repeated experiment with material objects.

Man can learn what are called "The Laws of Nature" by watching how similar objects behave under similar circumstances; and by repeating the experiment he confirms his certitude that the process is invariable.

In making these investigations, man confines himself to what is measurable. To deal only with what is measurable, to make the measurements accurate, to confirm them by repetition, is called "The Scientific Method"—whether it is applied to chemical phenomena, or to astronomy, or to archaeology, or to documents. So long as you are dealing with a material object which can be measured and with the reactions of material objects among themselves you are practising what is called in modern language "Science." But it is most important to remember that when you are dealing with other things, which cannot be subject to such a process, which cannot be exactly measured, which cannot be experimented with continually under identical conditions, the scientific method does not apply. The pretence that it does and that in these matters we can arrive through it at the same sort of certitude we have on physical laws is nonsense. For instance, the scientific method confirms you in the truth that certain musical notes are the product of repeated intermittent action called vibrations and vaguely spoken of as "waves." The scientific method can measure these in the case of music and in the case of colours. But it is quite worthless for the discovery of what music or what painting is beautiful. Physical science, arrived at by repeated experiment, man has always possessed—so far as we have any record of him. Among other forms of search for truth physical science has this particular importance, that by it man has to some extent acquired the power of controlling his material surroundings.

He achieves this mastery by the use of instruments which science enables him to produce. He produces them by combining various forms of scientific knowledge. These instruments we call tools and weapons. In their more complicated form we call them machines. But it is important to remember that there is no essential difference between the simplest instrument, such as a saw, and the most complicated piece of modern machinery, such as a motor-car. They are all of them the products of combining the results of experiment and observation in physical affairs.

The number and capacity of such instruments, if man be left with the opportunity to add to his knowledge, naturally increases with the process of time. But we must further remember that he has never had continuous enjoyment of such opportunities. There have been set-backs as well as advances. When there have been set-backs in the process there is a decline in the number and capacity of the instruments which man can use. Such loss occurs not only by wars, plagues, and natural catastrophes, but also by fatigue and by a change in the objects men set before themselves. Men may change from a mood in which they desire more and better instruments into a mood when they desire something quite different, so that the search for new instruments, and even the capacity for continuing to make the old ones, diminishes.

There are plenty of examples in history of big jumps forward in this respect, also long periods of neither advance nor retreat, and other periods of decline. There must have been a big jump producing all the main tools of carpentering, sculpture, and building, a jump which took place long before our earliest records. There was clearly a decline which began in our part of the world about 1,700 years ago, and then there was a long period of many centuries when things were more or less stable, without advance or retreat, and the same instruments were used from generation to generation.

The time in which we live is the latest phase. Perhaps the climax—and quite possibly the end—of a very big jump of this kind. And the characteristic of that time in our part of the world is a great development in the highly differentiated instruments we call machines, and side by side with them a great development of applied scientific knowledge.

This change has powerfully affected the life of our generation. Within living memory applied science has transformed great fields of social and individual action, and if we extend our time limit to a little more than a century its action is still more apparent. Almost all our instruments for transport—whether of men, goods, or ideas—many of our instruments for fashioning material, the great bulk of our weapons, and a considerable proportion of the things we use in daily habit have been profoundly affected by this change.

In the presence of such a disturbance all men are moved to ask themselves certain questions. These questions are often put confusedly and the answers to them ill thought out, but they can all be resolved into two main questions to our generation. Unless a right answer is reached to each of them we shall suffer. These two questions are : First, Is the possession of a new instrument a good in itself ? Second, How far are we controlled by instruments : are they our masters or are we masters of them?

Of these two questions the answer to the first one ought to be self-evident. The presence of a new instrument is in itself neither good nor bad. The only good or bad about the business is the use we make of that instrument.

Take a simple and fundamental case. It was found scientifically by experiment and thus established by proof that iron, if fashioned as a thin blade, could be given a sharp edge by rubbing it against certain other substances. It was found by experiment that iron grew soft when it was heated and got hard again when it became cool. It was found by experiment that if you hit a soft thing with a harder thing you can change its shape. By the combination of these pieces of scientific knowledge men got the instruments called the knife and the sword. Man had produced these novel things by the use of science, but they lay there before him neither good nor evil. He might use them for good or for evil, and it depended upon his mind which he did. He could with sharp iron fashion wood for a shelter against the weather, or cut another man's throat in a fit of bad temper, or his own in a fit of depression. What has been true of the knife for we know not how many thousands of years is true to-day of the flying machine or the latest explosive or the most recent poisonous chemical.

It is neither good nor evil in itself, the good or the evil resides entirely in its use, and that use resides in the intention of man. The mind governs.

I have said that this should be self-evident, but it must be repeated at the outset because in the moral chaos of our time and the absence of a fixed set of principles a great and perhaps increasing number of people talk as though an increase in the number and capacity of instruments was a good thing in itself.

There is in this connection one smaller point to be noted which is often overlooked, namely that every instrument has attached to it disadvantage as well as advantage inherent in its action. For instance, a new machine which enables you to cross the Atlantic in five days may make the passenger suffer much more from vibration than an old one which did it in twice the time. And if on balance the disadvantage is greater than the advantage, then to use the machine is not a good but an evil.

But it is the second question which is the more important and certainly the one which is now most disturbing and most continually occupying the modern mind. How far are instruments our masters? Or how far are we masters of them?

It is quite obvious that in some degree every new instrument, if its use be permitted, will affect human life. We say, talking loosely, that the invention of the plough turned men from pastoral to agricultural communities. We say that the invention of the railway both created great cities and compelled men to the new form of travel. I repeat, the phrases are loose and metaphorical, but in practice they will serve, for in point of fact a new instrument providing some good leads men into the habit of its use, and that habit produces a network of other connected habits by which man is in some degree controlled.

But in what degree? Everything lies in the answer to that question, and, indeed, all the more important questions set for mankind depend upon this point of degree.

What we have to determine is not whether machines in part control mankind. Of course they do. Nor whether we may not on occasions subject them to our will. Of course we can. Those men must be rare indeed in England to-day who are not continually compelled to travel by the aid of petrol; but those men are just as rare who cannot decide on particular occasions whether they will travel thus or no. The essential of the problem is to discover where the balance lies. Does the initiative lie *mainly* with us, with our wills, as individuals and as groups of individuals, or are we *in the main* the passive subjects of blind forces which our own activities have let loose?

Upon the right answer to that continual question turns the philosophy of our time, its general mood, and the happiness or unhappiness of mankind.

Now the answer which you hear most commonly in this country, which is given almost universally in great sections of society and which is widely heard everywhere, is that we are controlled by these things we have ourselves created. The initiative remaining to us is a minor factor in the business as a whole, the effect of the instrument upon us is the major factor. Is that answer the right one?

At bottom the discussion is simply the old discussion which dates from immemorial time between destiny and free-will. If the general answer comes to be given permanently against free-will, one type of society will result. If the answer is given the other way, another and almost opposite type of society will result. Of such moment is the debate.

The conviction against free-will is reinforced by propositions which pretend to be scientific and which would establish as a fact based on proof and admitting of no denial that human action follows upon forces extraneous to the will. But these affirmations are not scientific. The experiments cannot of their nature be identical or continuous or universal. We are all conscious within ourselves of the action of the will, and if we call it an illusion we do so because we have accepted a certain mentality, philosophy, or mood—not because we have reluctantly admitted the case proved against us by experiment. To think otherwise is to put the cart before the horse, for it is historically certain that the conviction of destiny and the denial of free-will came long before the insufficient and inconclusive experiments which pretend to decide the matter by observation. Experiment and observation are brought in here to confirm what true or false philosophy had already concluded, and not the other way about.

The belief that man is controlled by his environment and not his environment by man is powerfully reinforced to-day by a mass of regulation and constraint, more widely spread in some societies than others, but evidently present in a higher degree throughout our civilization than it was, even a generation ago, and far more than it was a lifetime ago. A uniform type of education is imposed by the State upon the mass of its citizens at a moment when their minds are being formed as children. In adult life every detail of action becomes more and more subject to external regulation, and under the pressure of such a political mood men naturally tend to the general philosophy that man is not a free agent; "the slave," said the wise man of antiquity, "thinks like a slave."

The feeling is further reinforced by an historical argument. We are told that the historical process has always been as follows: first a new material environment; then a change in the mind of man effected by that environment. Thus we are told that the invention of the printing-press was the main force in producing both the Renaissance and the Reformation. We were told not so long ago that the use of steam in travel would weld men together into one nation. Now we are told that instantaneous communication of ideas by telephone and telegraphy and far more rapid transit than steam ever gave have just the opposite effect and make human enmities more bitter than ever. The two conclusions are contradictory, but they spring from the same source. Each takes it for granted that the machine is the master of man.

In this great debate, the fundamental debate of our time, the arguments upon the other side are less often heard. It is with these I would conclude.

There are two kinds, the one drawn from observation and therefore themselves essentially scientific; the others pragmatic—that is, drawn from the consideration of consequences, relying upon the results that would follow in practice if the false philosophy were to be adopted.

The arguments from observation, the strictly scientific arguments, against the false and only so-called scientific conclusion that the instrument is the master of man is simply this: that if you look about you, if you concern yourself with the actual evidence and not with the guesswork or hypothesis, the evidence is against the constraint of man by machinery. It is against the thesis that man is the creature of his environment. This you can see in two ways. First, that the great mass of restriction to which man is subjected in the states which suffer most from such things is in no way the result of any modern scientific development but wholly political. Secondly, from the equally plain evidence that the degree of restriction varies very greatly between different countries and that the variation has nothing to do with scientific attainment. Take sumptuary laws. In one nation the citizen is prevented from the free use of wine or beer by force. In another he may only use these things in public places during certain restricted hours. In another, for instance in Germany, he has almost complete freedom. The variation is not a function of scientific attainment, it is a function of political mood. The same is true of the dead mechanical level of State education. The same is true of regulations forbidding men to work in their own shops after a certain hour. The same is true of any one of the thousands of constraints which had been imagined for the enregimentation and external forcing of human life.

Again, it is not historically true that the instrument preceded the mood. Capitalism, for instance, had already been established before modern machinery came in to serve it, and that machinery might just as well have served a different form of society. It is not true that the great movements called the Renaissance and the Reformation proceeded from the capital invention of printing. They used it when it came, but their origins were prior to its coming. The mood of the Reformers, the mood of the Renaissance scholars and artists, was earlier by two or three lifetimes than the mechanical changes at the end of the Middle Ages. So far as evidence and reasoning go the argument is all on that side. Instruments affect men, but man can control them, and his destiny depends ultimately not upon the dead object he himself has framed but on the attitude of himself.

Now when we take in its last place the practical argument I find it the most conclusive. But even for those for whom it may have no intellectual value it must have a political value.

It is this. If we do not exercise our freedom of choice, if we do not react, as we are capable of reacting, against the uniformity of a mechanical civilization, then we perish.

A mechanical civilization is almost a contradiction in terms unless we give the word civilization the mere meaning of "state of society." A mechanical civilization or culture in the sense in which we talk of the civilization or culture of Rome and Greece, France and England, Byzantium and Venice, is actually a contradiction in terms. Their traditions of beauty in building, in letters and the plastic arts, their tradition of debate in philosophy and religion, their whole body of multiple thought and achievement—detest mechanical rigidity. To be mechanical is to cease to be civilized. And for this reason, that the culture, the fruition, the happiness of society, its possession of the living soul, depends upon the faculty of choice in man. The very essential of life is multiplicity, and variety proceeding from the manifold human spirit is that one necessary factor without which a human society ceases to be and according to the degree of which it is a higher or a baser society.

We have before us in this considerable modern crisis—this conflict between the machine and the man—a plain duty, which is to use our wills everywhere for the defence of Will. To make it our choice to invigorate and multiply choice.

We must guard what is left of our freedom and extend it, we must fight collective control, we must mistrust the expert, we must question restriction wherever it appears, compelling it to prove itself necessary (as clearly it must be in particular cases), throwing the weight of proof upon the enemies of liberty and taking the rights of individual selection for granted. We must, in the economic sphere, fight not for greater collectivity of property, but, on the contrary, for greater distribution of it; we must fight for the small unit against the large one, for the self-governing guild against the merger and the combine. And these things we must do because the opposite policy—that which we have all been pursuing too long—leads rapidly to death.

J. B. S. HALDANE

5. THE BIOLOGIST AND SOCIETY

FIRST I want to defend my own profession. Mr. Belloc has said that we scientists can only deal with what is measurable. That is not the case. It is a scientific fact that hydrogen sulphide has a bad smell in the opinion of ninety-nine people out of a hundred. Now a smell is a sensation, something in the mind. You cannot weigh it or measure it. So far science can only deal accurately with rather simple things in the mind, like smells and colours. But that simple example shows that science does not deal only with material objects, but with the mind that knows them.

Mr. Hugh Fausset declared that the scientific approach to reality was one-sided, and that we scientists tried to reduce our minds to machines. Now of course there are one-sided scientists, just as there are one-sided painters and poets. But every scientific man or woman, at least in the experimental sciences, must combine three qualities: reason, imagination, and manual skill. And these must be welded together by an almost passionate acceptance of reality. Let us compare a great artist, Blake, and a great scientist, Faraday. Blake saw angels and devils where no one else could see them; he tried to fit them into an intellectual system; and he drew them and wrote about them extremely well. Faraday, who was a man of original and flaming imagination, saw lines of force where everyone before him had seen empty space. He reasoned about their properties. He devised and made apparatus to test these properties. And nine times out of ten, so he tells us, his imagination proved wrong. But his feeling for reality—his love of truth, if you like—was stronger than his imagination. He did not publish his imaginings unless they conformed to reality wherever he could test that conformity. As a result of Faraday's work you are able to listen to the wireless. But more than that, as a result of Faraday's work scientifically educated men and women have an altogether richer view of the world: for them, apparently empty space is full of the most intricate and beautiful patterns. So Faraday, just because he was a more complete man, as I think, than Blake, gave the world not only fresh wealth but fresh beauty.

But I must leave physics to Sir Oliver Lodge. I want to talk about my own science of biology—the study of living beings, including men. When Mr. Belloc talked about the progress of science in the last few centuries he was mainly concerned with machinery, and said, quite truly, that a machine was good or bad according to how we use it. Now beside the great physical and chemical discoveries there are great biological discoveries, especially in the field of medicine, which Mr. Belloc did not mention. We have found out how, by organizing the water supply, we can stop great plagues such as typhoid and cholera. Now that kind of discovery, provided it is used at all, is an almost unmixed benefit. Nobody, outside a short story, has ever used it to make an epidemic of cholera or typhoid, and I doubt whether they could.

We are only at the beginning of medicine, and have very little idea how much farther it may be going. That depends not only on the progress of medical science, but on whether the average man and woman can be got to think biologically. Diphtheria and scarlet fever could be made as rare as typhoid if people could be induced to take the necessary quite simple precautions. If politicians thought in terms of biology as they now think in terms of economics they would realize that it is at least as important to keep out foreign diseases as foreign imports. I do not think this is a hopeless ideal. Ordinary people are beginning to think to-day as scientists were thinking a hundred years ago, in terms of physics. They understand what is meant by such words as voltage and self-induction. A hundred years hence biological ideas may be equally familiar. In that case there will be an end of the tendency—of which Mr. Belloc has written—to put the machine before the man.

Now I want to suggest what would be happening to-day if this nation and other nations were biologically minded. The reasons for the present crisis are fairly simple, and the most important is this: Science has immensely increased our capacity for production. No attempt has been made to ensure that the goods so produced could be distributed. I suspect that this is partly the fault of capitalism, but we certainly cannot blame capitalism alone. We must blame the fact that, in our public thinking, we have never considered the results of the prolongation of life which we owe to medicine. Not only do many men live beyond 60, but they preserve a great deal of vigour. The average age of the Cabinet in March 1932 was 57. None of them were under 40, and 9 of the 20 were over 60. In a former great national crisis, 150 years ago, William Pitt became Prime Minister at 25.

Now a man of 60 or over has generally gained a lot of experience of life. But he cannot adapt himself easily to a new situation, and the present world situation is something entirely new. You cannot expect old men to deal with it. The Universities of Cambridge and Oxford are supposed to be very backward and reactionary institutions, but they have just agreed to a rule by which professors must retire at 65, not because their intellects have decayed, but because, with rare exceptions, they cannot keep pace with the rapid increase of knowledge. Not only our politics, but our industry, is controlled by old men. The average age of the directors of a number of representative companies is 62. Neither capitalism nor any other economic system could keep abreast of the times under such guidance.

An electorate which thought in terms of human biology would see that at least a third of the Cabinet were under 40, and not more than a third over 60. They would also take measures to transfer the control of industry to younger men, whether its ownership was public or private.

To-day we are engaged in imposing tariffs on a number of foreign imports, and members of one great party believe that this will encourage British industries. But the arguments for and against tariffs are wholly almost economic. The biological side of the case is quite neglected. Now what would happen if any members of the Cabinet took human biology seriously? They would consider the effect of encouraging any particular industry on the national health. A tax on imports of cut flowers will improve the national health, because it will encourage more men to enter the very healthy occupation of gardening. If you take ten thousand gardeners and ten thousand men of the same ages from other occupations, you will find that at the end of a year 432 from the general population and only 305 gardeners will have died. So by encouraging gardening you make Britain more healthy. But now suppose the question came up of a tariff designed to encourage the manufacture of files. The death-rate among file-cutters is 85 per cent above the average—nearly double. No, our biologist-politicians would say, we cannot protect your industry until you make it reasonably healthy. Install proper protection against poisoning your workers with dust, and when you bring your death-rate down you shall have your tariff—but not till then. It is just the same with scores of other questions which are now being argued on economic lines. The moment you take the biological point of view, you start thinking of the man before the machine, of health before wealth.

Now what would be the international policy of a Cabinet who looked at world affairs from the biological, not the economic, angle? I believe that they would consult with foreign colleagues of similar temper, and begin preparations for the next world war. Yes, the next world war. War is a terrible thing, but it satisfies certain deep-seated desires in many men, including myself. If we cannot satisfy these desires somehow else, we shall do it by killing our fellows as we did in the last great war. But I believe they can be satisfied in other ways. The war of which I am speaking is a war "not against flesh and blood, but against principalities, against powers, against the rulers of the darkness of this world," a war against the agencies of disease. If mankind co-operated, we could abolish for ever a whole group of pestilences, such as smallpox, cholera, typhoid, diphtheria, and scarlet fever, and such carriers of pestilence as the louse. A campaign of this kind would involve the co-operation of the whole world. As long as a single Indian is suffering from cholera, a single negro in Central Africa from smallpox, there is a focus for that disease from which you or I may be infected. I picture a great army of men and women sweeping across each continent as the armies swept across Europe in the Great War, systematically sterilizing every human being and every house and leaving a clean world behind them. Infectious territories would be blockaded as Germany was blockaded in the Great War. There would be hardships and dangers. But men and women welcome hardship and danger in a great cause.

If you are to have real world co-operation and world loyalty, it must be co-operation for some positive end, not for a merely negative end, such as keeping peace. Now I expect most of you think that the idea of a world war against disease is a silly idea. It is a silly idea now. Fifty years ago the idea of a regular air mail service between England and Australia would have been a silly idea. To-day it is a working idea. Such ideas as I am putting forward will be working ideas fifty years hence if enough people want them to be.

Medicine is at present the most important branch of human biology, but it will not always be so. In the first place some people are born with such great physical or mental handicaps that no amount of treatment can make them normal. Secondly, health, whether of body or mind, is only one of many things needed for a good life. It would undoubtedly be possible to prevent the birth of a great many of the defectives, though not perhaps quite so easy as Mr. Aldous Huxley seems to think. But to make the average of the population cleverer or—as Mr. Huxley suggests—stupider, much less to breed men of genius, would be an altogether harder task. Let me tell you why. For twenty-seven years a set of quite competent biologists have been studying inheritance in a particular species of primrose which has a generation each year. So those twenty-seven years correspond to about seven hundred years of human history. Although they could breed these plants as they liked, some combinations of shapes and colours have only just been made for the first time. I have here one of our new combinations which we made this year for the first time. I wish you could see it. It has curious twisted petals with toothed edges. It is like one of Mr. Aldous Huxley's novels: either you like it very much, or you think it rather unpleasant. I like it, but I expect about half of you would hate it. Well, seven hundred years hence we may know enough of human genetics to say that the child of two particular parents will probably be a writer of the type of Mr. Aldous Huxley. I doubt if we shall get so far in a much shorter time. But long before that time the study of human genetics will have forced us to realize that, with a very few exceptions, every human being is born quite different from any other, that nothing we can do to them will make them alike, and that it is no good trying. As soon as that elementary fact of human biology is realized, not as a mere abstract statement, but as a scientific law supported by innumerable detailed examples, I think that we may look forward to a rebirth of individual liberty. Bertrand Russell and Mr. Aldous Huxley have tried to make your flesh creep with prophecies about the future tyranny of the expert, who is going to do his best to make men uniform, like factory-made goods. That would be plausible enough if physics were the only science. But clearly a great many of the experts would have to be experts in biology, as doctors are to-day. Now one of the first things a biologist learns is that no two frogs, let alone two men, are quite alike. And if he is a geneticist, studying heredity, he is mostly concerned with those differences, and particularly with the odd individuals who keep cropping up when we think we have got a true-breeding population. The founder of genetics in this country, William Bateson, left a motto for his successors, of whom I am one. It was "Treasure your exceptions." Treasure your exceptions. That represents the biological point of view, whereas the engineer's point of view is "Scrap your exceptions." In so far as our expert rulers of the future are biologists, I am sure they will be a great deal more tolerant than our present rulers of human exceptions, the men and women who do not fit into society as at present constituted, sometimes because they are too bad for it, but occasionally because they are too good for it.

A society based on the recognition of human diversity might adopt a eugenic programme, but it would at any rate make the nature of man, not that of machines or institutions, the foundation of its social philosophy. To a biologist the social problem is not "How can we get these men and women to fit into society?" but "How can we make a society into which these men and women will fit?" Our present attitude is quite different. For example, reformers are constantly putting forward educational schemes, such as the Dalton plan. Some boys and girls learn far better under these new methods than under the old one. But I expect the opposite is also true. The old-fashioned methods worked fairly well on me, and I found the sort of education that was thought up-to-date thirty years ago very boring. For any individual child there is some ideal system of education, and educational methods will not be perfected until we discover how to find that out in any given case.

The same is true with the choice of a career. I am going to quote from Professor Spearman, a psychologist who is so scientific that a special branch of mathematics has had to be created to enable him to investigate the human mind. This is what he says: "Every normal man, woman, or child is a genius at something, as well as an idiot at something. It remains to discover what—at any rate in respect of the genius." No one has done more than he to find methods for discovering the nature of individual genius.

In our present society industry is not even organized so as to find a job for every able and willing citizen. That will have to be done if civilization is not to collapse. But in a scientific society the attempt would be made to find the best possible jobs for everybody. This is how the ruling group of a scientifically organized society would be thinking three hundred years hence: "The psychologists who have been testing the school-children tell us that we may expect three or four really first-rate musical composers ten years hence. That will mean building at least one new broadcasting station. But there is a serious shortage of boys and girls who are likely to be really first-rate air pilots. Probably only 15 per cent of our people will be able to fly to work in the morning, instead of 25 per cent, as we had hoped. We shall have to modify our town-planning schemes, and put down a number of our new factories in the country, instead of bringing the workers in every morning." That is the way people would think in a civilization in which machines were regarded as secondary to men.

Now for the question of liberty. I think it is fairly clear that we shall have to sacrifice a certain amount of our present liberties in the economic field. Either industry will be organized more and more in great corporations, private or public, or else legislation will be needed to prevent this tendency. I think that if centralization is accompanied by a measure of control by the workers in it, the net result would be an increase of liberty. Again I believe there will be a progressive decrease of what Mr. Belloc calls parental liberty, but what, from the child's point of view, is often parental despotism. The liberty to flog your child daily and lock it up in a dark room is like the liberty of a medieval baron to hang his serfs. It is a very one-sided sort of liberty indeed. Incidentally, one reason why the biological point of view is rare is that parents often protest when human biology is taught to their children. There will probably be a decrease of liberty as regards reproduction, where it is highly probable that any child born to a couple would be a victim of serious physical or mental defect. There may be a revival, on eugenic grounds, of the medieval restriction on marriage of first cousins. But there would be a corresponding removal of existing restrictions to love and marriage which are not based on ascertainable fact. There will probably be more regulations, but regulations may make for more, not less, liberty. I can drive my car as fast as it will go on the open road, but if I am to do this I must conform to the road code, and carry licences and other documents. This seems to me fair enough. Such regulations have made for increased freedom on the road for all motorists except road-hogs and bandits. I think they are inevitable as the material basis of our civilization becomes more complex. The introduction of radio has made a number of regulations inevitable. But it has not diminished freedom.

Some of our existing unnecessary regulations are

based on tradition. Most of them are based on the great principle that if a few people misuse something, the rest must be deprived of it. Such are our laws about liquor, and the even more ridiculous law which prohibits children from smoking. Now it is a generally recognized biological law that one man's meat is another man's poison. But few people realize that it follows that one man's poison is another man's meat. A scientific civilization would recognize that there are a minority of people who cannot be trusted with liquor, just as others cannot be trusted with a car, and it would probably treat them in much the same way. If I could get a drink whenever I wanted it, I should be quite willing to carry a drinking licence about, knowing that it would be endorsed or taken away if I got drunk.

I cannot agree with Mr. Belloc that the philosophical theory of the freedom of the will has much to do with the question. You cannot make a sufficiently good man do what he believes to be wrong by any threats or bribes. In that sense the will is free. But you can, in some cases, find out the reasons why some men are better than others. If that were not so, we could not help our fellows to be good. With the aid of psychology and genetics we can go a long way to analyse the springs of character. In so far as this analysis is possible the will is determined by known causes.

But because most biologists do not believe in free-will in Mr. Belloc's sense, please do not suppose that they believe in Mr. Huxley's tall stories about the possibilities of moulding the character from outside. I do not claim to know much psychology, but I know that the Behaviourists, an American group of psychologists who claim that they can do what they like with children's behaviour, are generally wrong in their physiology, where I *can* check them.

Four distinguished writers, one of whom is also a very distinguished mathematician, have already contributed to this symposium. All were mainly concerned with applied science, and not with the results which would follow if the average man and woman adopted a scientific outlook. That is natural enough. They learned their science from books, and books usually present science as a finished product, not something alive and growing, as it is in the mind of the scientist. If you really understand how your receiver works, not from books, but from experimenting with its parts, you have got a more scientific outlook than you will get from a dozen books. When anyone tells you something about science, ask yourselves if he or she has only learnt science from books, or with their hands as well. Many of you are skilled manual workers. So are most scientific men and women, including Sir Oliver Lodge, who is making the final contribution. This is one reason why their point of view is in some ways more human than the politician's or the writer's. The scientist in his laboratory has to deal with *this* crystal or *this* flower, not with crystals and flowers in general. If the rulers of this country ever adopt the scientific point of view they will realize that they are controlling the destinies of forty million men and women, all individual, and all unique. And men are civilized just in so far as they realize the uniqueness of their fellows, and act upon their knowledge.

OLIVER LODGE

6. THE SPIRIT OF SCIENCE

I HAVE been impressed by the amount of agreement which one can feel with many of the preceding contributions to this symposium. There are points of difference, of course. I do not agree with all Bertrand Russell's outlook; in fact, I often seriously disagree with him, but on this occasion less than usual; and, surprisingly, I am able to agree with much that Mr. Hilaire Belloc said; but then he dealt mainly not with science but with the applications of science, which he summed up under the term "machines." These are made possible by science, but the responsibility for their use or abuse belongs not to science but to civilization. If so-called civilization allows machinery to sap human freedom and enslave mankind, science washes its hands of any such egregious folly. That human welfare is the first thing to aim at, and that of all industrial products or national manufactures the production of human souls of good quality is the most ultimately remunerative, "quite leadingly lucrative," has been said forcibly and eloquently by John Ruskin. Let us attend to his teachings in that at one time heretic pronouncement *Unto This Last*.

Mr. Aldous Huxley is always interesting and usually provocative, and on this occasion his exposition of the methods and powers of science, and his distinction between the aims of the economist and the humanist in its modes of application, are well worthy of attention. Not till he sets forth the preposterous claims of the extreme Behaviourist does he, perhaps purposely, lose touch with common sense; though he returns to it in his conclusion.

With what Mr. Fausset said I am in considerable agreement, as will be seen by what follows. While of most of Mr. Haldane's contribution I can only express respectful admiration. The scope of Biology at the present time, especially in the new departments of Bio-chemistry and Bio-physics, is intensely interesting and of the utmost importance; and of this great study Mr. Haldane is a distinguished representative. With what he has said in criticism of the previous speakers I agree in general, and I prefer to dwell upon the parts where we are all in agreement; as will doubtless appear in what I have to say on my own account, except that in some cases I may be disposed to go farther than the rest.

Where I differ from Mr. Haldane is where he touches on what he might call branches of " human biology," namely the sciences of Economics and Eugenics. I doubt if electors and politicians, however biologically minded, will ever take the view that youth and enterprise are more to be trusted than age and experience in the conduct of affairs. Surely there is room for the aptitudes of both the old and the young. As to the discriminative use of artificial modes of stimulating or protecting various kinds of industry, healthy conditions for workers could surely be investigated and supplied without the reward of a tariff. My private hope indeed is that custom-houses all over the world will be abolished, and an era of free interchange of goods established, before mankind is one century older. The present move in the direction of assisting empire trade may, I trust, be an indirect step towards the ultimate attainment of that desirable end.

The science of Economics seems to demand much wisdom in its application. It has absorbed a great deal of attention, but I do not feel sure how far it is to be trusted; and the wisdom has often been lacking. The younger science of Eugenics has not yet advanced far enough for us to contemplate the breeding of genius or of any specific brand of human being. I doubt if our knowledge can ever be wisely applied in that direction. If we interfere with humanity to that extent we are trying to play the part of Providence, and may make a mess of it. As to the manufacture of human beings in a laboratory or in a bottle, that, in spite of the author of *Daedalus*, seems to me a lunatic idea. We have not yet vivified a single cell; though indeed that achievement does seem a possibility for the remote future.

I mistrust the application of Eugenics to humanity in the positive direction: it may be necessary and legitimate to eliminate the unfit, but not to decide on any particular brand of human stock. Uniformity is not a thing to aim at, "It takes many sorts to make a world," as the proverb says. Let us accept what Mr. Haldane tells us is true, that no two animals are exactly alike, and let us remember Bateson's slogan, "Treasure your exceptions." Attention to unexpected and exceptional results is the mode in which every scientific man is led to surprising discoveries of real novelty; and many have been missed by neglecting that attention. Every exception is full of instruction and may be of great value. There is no mould into which human beings can wisely be fitted, they must be free to develop idiosyncrasies. A civilized community would give individuals a chance of developing their latent powers, and then, so long as they were healthy in body and mind, it would leave them free.

It seems to be a question whether the spread of science is good for civilization or not, or in what way it is good and in what way it does harm. Let us see what we mean by science: I mean science itself, not its applications. Many definitions have been given, some of them too narrow, like "Science is measurement," which reminds one of Eddington's "Pointer readings"; others, perhaps too general, like the excellent phrase "Organized common sense," used by Thomas Henry Huxley. If I were asked for a definition of science I might say: "Accurate knowledge concerning things which appeal to the senses and concerning everything that can be inferred about the causes of their behaviour." Or I might modify that, because things that appeal to the senses are literally "phenomena" —that is, appearances—so the definition of science is equivalent to this: Accurate knowledge of phenomena, and inferences about the reality which underlies them. This assumes that there is a reality underlying all phenomena. There must be some reality to account for the appearance. The appearance might be a will-o'-the-wisp, or an image in a looking-glass, but still full knowledge of it will enable us to say to what it is due; and it may be explained either chemically, in the one case, or optically, in the other.

Accurate knowledge does not necessarily mean metrical knowledge: it includes measurement, but many things which are the subject of scientific inquiry are not metrical, especially a number of the facts studied in Biology. These are not amenable to calculation, and yet our knowledge of them may be accurate.

Many of the definitions that have been given of science tend unduly to limit it, and do not lay sufficient stress upon the scope of *inference*. Practically everything may be made the object of scientific inquiry. Palaeography, for instance, or the scrutiny of ancient manuscripts: they can be examined and interpreted scientifically. Psychology, or a study of the human mind, is an undoubted science; though the manner of exploration differs in many respects from customary scientific procedure. Even Theology may be treated from the scientific point of view, and has been called "the Queen of the Sciences," though the subject-matter there is infinite, and mainly beyond our comprehension.

The things which are often popularly understood under the heading "science" are really only the applications of scientific knowledge to human purposes. These applications belong more to civilization than to science, they are more of the nature of engineering. Engineering is undoubtedly based on science; but the constructions that it makes, and that humanity uses, are the result of organized labour. It may be a miracle of design, like the Forth Bridge; but the uses to which it is put depend on the will of the community. The same is true of a machine. A machine is designed by an engineer with scientific knowledge; it is constructed according to plan by a number of workmen; and it is used in such way as the community may decide, either for good, neutral, or evil purposes. Everything may be used, or may be abused. In the last resort the owner, or the community, has to decide. The mind of man is ultimately the important thing; and that means the nature and amount of civilization.

"Civilization" must mean the art of living together as a community, a variety of what biologists call symbiosis. In a good civilization the action of every part contributes to the welfare of the whole. The community and the individual then thrive together. If sections proceed to fight and seek to destroy each other, that is not a result of scientific knowledge, but of defective civilization. If our civilization is full of evil tendencies of this kind, and aims at destruction, scientific knowledge makes it more dangerous, and gives it powers for evil which it otherwise would not possess: but it can always exert itself for destruction up to the limits of its power, if it is so inclined. The will, both of the individual and of the community, is the main thing to get right, if we are to live together in a civilized manner. Greed and selfishness are the root of all evil, and scientific weapons may increase the powers of evilly disposed persons. In some excited states of society people have attended more to powers of destruction than to powers of beneficence, and have been willing to expend their corporate savings in that wasteful and deadly way. We know by bitter experience what the result of that is. One of the results is the present crisis in which the world finds itself.

Mr. Haldane spoke airily of the next "war"; but he went on to explain his meaning. It was not an internecine warfare of sections of humanity against each other. It was an international corporate warfare, marching shoulder to shoulder against remediable evils, such as plague, pestilence, and famine. It was a war, he said, not against humanity, but "against principalities, against powers, against the rulers of the darkness of this world," a wrestling with the uncivilized unconscious powers of destruction, which still exert their deleterious influence; a war against diseases and social evils of every kind, and the agents which propagate them.

Biologists, led by Pasteur, have now ascertained the causes of many diseases, and some biologists have shown the way, and been assisted by the community, to take measures to stamp them out. Thus it was when the United States of America sent the engineer Colonel Gorgas to make habitable the region near Panama, where white men had suffered and died like flies under the ravages of yellow fever and malaria. It was an administrative operation, applying the results of scientific discovery. Ross and others had discovered the germs of malaria in the mosquito, and pointed out that the way to attack the disease was to destroy and stop the breeding of the noxious insect; and by administrative regulations the task was accomplished. In another field, another great biologist, Sir David Bruce, set to work, travelling about in Africa in order to ascertain the causes, and thus be enabled to take measures for the prevention, of sleeping-sickness, a deadly disease conveyed by the trypanosomes carried by the tsetse fly. When that fact was discovered there was still much to be done. Bruce secured the co-operation of native chiefs, who sent out search-parties in different districts to find out the habits and localities infested by these creatures. He was only able to make a beginning— Africa is a large continent. The Governments must be interested in the problem, and must give facilities for wholesale treatment. The way has been pointed out by science: it remains for the will of humanity to do the rest.

There are other scourges afflicting man, who thinks he is civilized, which have still to be tackled. At present they seem too much regarded as inevitable, and not worth an effort. Effort is thought to be hopeless. But if the corporate will of humanity were directed to that end, if bands of scientific explorers were encouraged to find a remedy, and if then steps were taken to apply the knowledge on a large scale, these foes of humanity could also be overcome. The two great demands on the good will and energy of mankind at the present time are: more science—that is, more organized knowledge; and more civilization, or the determination to apply that knowledge in good and beneficent directions.

Science and civilization should co-operate, as a pair of partners. Each can strengthen and supplement the other; and together, jointly, they can attain results impossible to either separate. Increase of knowledge is all to the good, but by itself is barren of results. Knowledge can fructify and bestir the community to apply it in beneficent directions, and to bring forth the fruits of true civilization. So long as we have knowledge which we do not apply to good purposes, we are not really civilized. The workers in science are few and enthusiastic. They sometimes lack the means even to carry out their exploration to a fruitful end. They certainly have no power of extensive application. The real need is, first of all, for the community to encourage the workers and provide the facilities, and next to have the good will to apply the discoveries on a large scale in beneficent directions.

Science and civilization are two forces that should work hand in hand and progress together. At present science has gone ahead, civilization has lagged behind. The only time when the community has really encouraged the scientific worker and made extensive use of his labours is when it was engaged on the uncivilized work of destruction, as in the late war. Then Government grants were forthcoming, expense seemed no object, and there were not wanting factories for the supply of munitions of all kinds. And still the nations feel a jealousy and a fear of each other, and maintain an expensive armament to attain what they imagine is security. That is not the way to attain security: it is attainable only by indirect methods. The will of mankind has to be set in the right direction. True civilization is the power of living together on this planet in mutual co-operation; the nations not squabbling and seeking to damage each other, but pooling their resources, combining their knowledge, and advancing with a united front against the real foes of humanity as a whole.

Youth is willing to serve in this enterprise. There

is excitement in it and danger. Many investigators have already succumbed to tropical diseases; and sometimes an investigator subjects himself to a disease in order to study its symptoms with greater precision. These are risks: and if there are scars, they are wounds received in a glorious war. This is the war of the future, to which the young of all nations are called. There is plenty of high spirit, at present lacking direction; plenty of energy, not knowing how to expend itself; many unemployed, who lack the power of initiating enterprises for themselves. Not so much *more* education is wanted as education in the right direction; more scope for individual exertion; less dependence on the enterprise of others; more initiative of our own. It is a striking piece of exaggeration to say that "everyone is a genius at something, and everyone is stupid at something." Our present system of education finds out the stupidity readily enough; but genius generally has to find itself out, in spite of the system. How to remedy defects in our educational methods which result in a genius being commonly thought stupid at school—that is beyond me, but it is not beyond the powers of humanity. If we encourage the spontaneity of youth, and liberate it in effective directions, a vast deal more could be accomplished. What is done in science is done mainly by youths who gradually become conscious of their own aptitude, and struggle into a position where they can develop it. The history of scientific discovery and invention is full of examples of hostile circumstances overcome by individual enterprise and energy. But how many there must be who fail to find the mountain path, and who sink back depressed into the general average.

Meanwhile, and in spite of everything, knowledge is advancing at a tremendous rate: not all of it so immediately applicable to human welfare as Biology is. But every increase of knowledge ought to advance true civilization, and enlarge our conception of the material universe, which is our temporary home, and in which we have opportunities for service. The revelations of Astronomy have enlarged the universe beyond all previous conception, and raised strange problems about time and space and the fundamental nature of matter and energy—problems which exercise the strongest brains among us and give much food for philosophic thought. All this study is valuable in the highest degree. It does not affect the bodily welfare of man, but it affects, what is more important, his whole faith and outlook. Man is a spirit, he does not live by material considerations alone, he can look before and after, and attain a wide comprehension of Reality. He is led by the evidence of his senses up to a certain point, and then has to depend on his powers of inference. With these he can soar beyond all sense-perception, and trace Reality in regions inaccessible to any sense-organ. All this too is or should become the region of science. It all conduces to a higher civilization. The universe contains far more than as yet we have any idea of. We are encompassed about with a spiritual world and are not yet aware of the fact, save from the reports of a few trespassers, to which neither science nor civilization attends. The Poet and Philosopher can expand into this region, and can receive inspiration for the still further elevation of humanity, until each generation as it passes can feel that

> Tho' from out our bourne of Time and Place
>   The flood may bear us far,
> We hope to see our Pilot face to face
>   When we have crost the bar.

INDEX

SCIENCE AT YOUR SERVICE

by

JULIAN S. HUXLEY, F.R.S.

SIR EDWARD APPLETON, K.C.B., F.R.S.

SIR GEORGE BURT

SIR LAWRENCE BRAGG, F.R.S.

PROFESSOR J. B. SPEAKMAN, D.Sc.

PROFESSOR JOHN READ, F.R.S.

DR. A. O. RANKINE, O.B.E., F.R.S.

SIR NELSON JOHNSON, K.C.B.

MICHAEL GRAHAM

DR. ALBERT PARKER

J. L. KENT

G. L. GROVES

E. C. BULLARD, F.R.S.

With a Preface by
E. C. BULLARD

Illustrated

London
GEORGE ALLEN & UNWIN LTD

CONTENTS

ILLUSTRATIONS

PREFACE

THIS book contains twelve talks broadcast by the B.B.C. in their Home Service during the winter of 1943-44. The series was called "Science at Your Service," and its object was to show the influence of science on everyday life. Each speaker dealt with some subject of practical importance with which he was specially acquainted, such as textiles, fish, or tunnels. Each talked about the difficulties of providing what people need, and explained how some of these difficulties have been overcome, and how the rest are being tackled.

These talks show that not all science is remote and theoretical, but that much of it affects our daily lives. Each speaker is talking about a subject to which he has devoted a large part of his energies for many years; each is therefore an enthusiast, and paints a rosy picture describing the successes that have been achieved, and the prospect of further successes in the future. Up to a point this optimism is justified. The application of science can increase the wealth of the world; but when all these talks are put together, as in this book, I think that in some ways they give a false impression of the social effects of science. They are too much like the advertisement of a patent medicine; in fact, too good to be true. One gets the feeling that science is a universal cure for all human ills, that it will restore our export trade, raise our standard of living, and provide everyone with ninepence for fourpence.

It is important to bear in mind that science does not do all these things automatically. It is just as capable of providing guns as butter; it can as easily be used to kill men as mosquitoes.

During the past three hundred years we have discovered how to make Nature give us what we want. The technique consists in using past experience to formulate general principles which we call "Laws of Nature," and applying these to decide what will happen in the future. The essential point is that it is experience that led us to rely on these principles, not intuition, or supernatural powers, or philosophical ideas. There is absolutely nothing in this process that ensures that the results are good or desirable. It is simply a process for securing a specified end.

The moral of this seems to me to be that science can only produce worth-while practical results if the social system is so contrived that applied science is directed to worth-while aims. No matter how many excellent scientists there had been in Nazi Germany the standard of living of the people would not have been improved. The only results would have been to accelerate the preparations for war and to make it more deadly.

Owing to their successes in this war scientists are, at the moment, popular people in this country; and there is, I think, a tendency to swing right over from an attitude that regarded science as unimportant to one that regards it as a cure for all ills. I am afraid that our hopes are likely to be disappointed unless we are a good deal more certain of what we want science to do for us than we were before the war. Our resources will always be limited, and there will always be the problem of choosing in what direction we want to use our time and efforts. Do we want more houses, cleaner milk, more aeroplanes, television, or better hospitals? We cannot have everything we want, but by proper application of science we can have a good deal. How shall we choose? There, perhaps, is the fundamental problem of politics.

From this point of view these talks are a kind of shop window. They show the kind of things we have and some of those we might have, and they give some idea of what sort of efforts are needed to get them. I think they will have achieved their purpose if they suggest to their listeners and readers how wide are the possibilities and how interesting the work waiting to be done when we can turn again from destruction to the works of peace.

E C BULLARD

LONDON

May 1944

SCIENCE AND THE HOUSE

By Julian S. Huxley, F.R.S.

Why should we need science to come butting in to our houses? Hasn't man been able without science to produce the most gracious and efficient homes? The answer is both yes and no.

It is true that a good Queen Anne house, for instance, is as attractive a place to live in as one could wish. But when it was built most of our population was still living in squalid hovels of insanitary if picturesque cottages. Indeed, throughout all history, most human beings have been badly or inadequately housed. Science is needed to help us remedy that.

A second reason is given by the particular situation in which we now find ourselves. The nineteenth century with its system of individualism and *laisser-faire* certainly produced wonderful progress in some directions, but in others its consequences were not so good. It produced the ugliest and dreariest as well as the biggest towns in history; new and larger slums; jerry-building; traffic chaos; ribbon development. The net result is so vast and so chaotic that nothing but drastic planning can possibly adjust it to real human and social needs; and planning, if it is to be any good, implies the use of scientific method.

The third reason is that in the long run science has something to say in every field of human life. As the house supplanted the hut and the primitive shelter, man gradually came to realize that he could, in his home, create an artificial environment for himself. In addition to the old needs of protection from heat and cold, rain and sun, and of security for person and property, humanity evolved new needs which houses had to meet—the need for comfort and convenience; for privacy at one time and sociability at another; for spaciousness; for beauty. The house could and should be a machine for living—and as with all machines it can be improved by science.

Sometimes science can improve something old and familiar. Thus it can test traditional materials like bricks, set up standards for their manufacture, find out their most economical use. Or it may give us things which are wholly new. Steel frame construction, ferro-concrete, gas cooking and electric lighting all represented quite novel advances and were all due to science. As an example from to-day, take lighting. It will soon be possible to instal a new type of daylight lamp as part of ordinary household equipment— it's in industrial use here already. This is a vacuum tube in which the discharge causes fluorescence in a specially prepared material. In this kind of lamp, much less of the energy required is wasted as heat and so there is much lower current consumption, though it's rather expensive to instal. Whether it will pay to instal it depends on further scientific improvements.

Then there is the quite recent invention of plastic adhesive or glue which is also waterproof. This is already employed in the construction of our Mosquito bombers. In building, its most important use will be to waterproof plywood. Plywood is itself a modern invention and if we could have it waterproof and damp-resistant, it would enormously increase its usefulness.

Plywood provides a specially good example of the use of scientific *method*. In the old days, any sheet of plywood which had a defect such as a knot, had to be rejected. But a knot won't matter unless it overlaps another knot in the next sheet. So the statisticians were then called in and asked to work out, on the theory of probability, how often this was likely to happen for different sized defects. On the basis of their work the percentage of rejection has decreased from around three-quarters to below half, and the actual output of certain plywoods has about doubled.

Let me go back to the waterproof plastic adhesive. Careful tests in Government Research Institutions, have shown that in an average house the outer wall of brick is somewhere about twice as thick as it need be for the support it has to give. So we could replace the inner half of the wall by something else which would give better heat insulation and so cut down fuel costs. One of the best materials is a synthetic boarding made from odd pieces of wood cut up fine.

Our new plastic could be made of any colour, applied to paper with a polished surface and then stuck on to the boarding. This gives a really lovely finish to a room, quite as attractive as the best distemper or wall paper; and it is cheap, washable and damp-resistant.

Other technical applications of science for housing have already been developed on a large scale in some other countries, such as the method of District Heating, by which heat is piped to all the buildings within an area, just as light or gas is now. Another is the Garchey system of refuse disposal for blocks of flats; all house refuse is tipped down a hole in your sink and then used, after treatment, as fuel to provide heat. We want such methods to be given a full and fair trial here.

Then science helps in another way by providing a firm basis for building standards. At the present moment various professional institutions concerned with building and architecture have set up a series of committees under the fatherly care of the Ministry of Works, and they're engaged in drafting Codes of Practice. These codes will incorporate all the best that the latest scientific knowledge can give, both as regards economy and efficiency of construction in all the different aspects of building. After they have been published—probably in a year or so—there will be no excuse for bad building, even though the codes are not likely to be made compulsory. So after the war, if you're thinking of moving into a new house, ask if it has been built in accordance with the new building codes.

In some cases, by the way, public authorities will have to change their regulations if we're to get the best results. Good sound insulation is essential as everyone knows who has lived in a block of flats where all sorts of harmless and even desirable activities are forbidden to tenants because of the noise they make. The cheapest and simplest method of sound insulation between two rooms is to build the party wall double with an empty space in the centre: but this is now prohibited under the by-laws covering building.

In general, these by-laws have been perfectly satisfactory and indeed essential safeguards for traditional methods of building, but sometimes they haven't provided adequately for new methods and materials, or have even hindered them. Such as the possibilities of reinforced concrete construction—as I found out over some buildings we wanted to put up at the Zoo.

But new materials and methods often also run up against popular prejudice. Thus, steel frame construction makes it possible to free the outer walls from the business of supporting a building, so that one can turn them into a mere curtain; even corners could be wholly made of glass. Now when the first building of this sort was put up in Paris, great difficulty was found in insuring it, as all the Insurance Companies were certain it was bound to fall down!

A similar example is provided by the beautiful bridges of that remarkable Swiss architect, Maillart, who pushed the principles of ferro-concrete construction in bridge-building to their logical conclusion. The resultant thin slabs and delicate arches looked so unsubstantial—though really just as strong as the most old-fashioned stone or steel bridge—that many local authorities in Switzerland refused to have anything to do with them.

New methods like steel-frame and ferro-concrete construction—so-called pre-fabrication, the manufacture of
parts of houses away from the site for assembly on the
building site—new materials like glass bricks and plastics—
all these are opening up quite new possibilities for architecture and building. We can, if we want to, indulge in
all sorts of new possibilities—new shapes, great expanses of
glass, movable partitions instead of party walls, new
furnishings and many other things impossible to the traditional methods of construction. But in order to enjoy these
new advantages, the public has got to get over their irra-
tional or purely traditional prejudices. They must accustom
themselves to the new possibilities of lightness and spaciousness, and of bringing the landscape to interpenetrate with
the interior of the house. It is a curious fact that though the
average man likes to have the latest possibilities devised
by science incorporated in his car, and would never think
of asking for one that looked like a 1900 model, in housing
he tends to be shy of novelty and will try to go back to
styles which were developed centuries ago, to meet the needs
of those past periods, rather than going forward to the
new style that is waiting to be born.

In any case, don't let's forget that beauty as well as
comfort is something which our houses can give us. It is
no more expensive to have good architecture and good
design in fittings and furnishings than it is to have bad
architecture—or no architecture at all—and shoddy, vulgar
design.

One field where public prejudice certainly helps to retard
advance is that of smoke abatement. Just think of it. About
two-thirds of our population lives perhaps two-thirds of
their lives under an artificial cloud, gloomy, dirty and
unhealthy. Though a great deal could be done about this
by regulations for industry, the bulk of this smoke-pall
comes from domestic fires. The open fire is an attractive
thing in the family living room, but if only the public would

accustom itself to use other forms of heating where possible, and at any rate insist upon smokeless fuel for the open grate, then we could go a long way towards healthier cities and more beautiful housing.

To come back to research, the Government Department of Scientific and Industrial Research founded in 1916 and its various branches, especially the Building Research Station and the National Physical Laboratory, have carried out an immense amount of research and testing on building problems and building materials.

Research—combined with new industrial techniques—has now given us the possibility of quite new standards of comfort, efficiency and beauty in our homes. But possibilities aren't actualities; and too often in the past, the knowledge provided by science has not been applied in practice. The most glaring example, I suppose, is the fact that long after we knew just what amount of different foodstuffs were needed for full health, something like a quarter of our population was still not properly fed.

To get these new possibilities realized in our houses, co-operation is needed from three different quarters—the builders and architects, the Government, and the public. The builders and architects are actively co-operating in all the Government schemes; and the public can always help in keeping them up to the mark. Further, the public—and that is all of us—must not let the immense need for quantity of housing which will face us after the war make us forget the equally great need for high quality in our houses. Otherwise, what will probably be the biggest housing programme in history will be unworthy of our country and its greatness.

But more demand won't give us what we want without the most careful planning and organization to carry it out. In particular, we need far-reaching Government decisions on various matters of principle. First on planning—the location of industry; the basis for compensation when land or development rights are bought for planning purposes; the grant of compulsory powers of acquiring land; and so on. Secondly, as regards financial assistance for good housing. During the war we have, in one important sphere of life, that of food, adopted the principle that what counts is not the ability to pay, but human needs. It is the babies and the mothers that get the oranges and the extra milk —not the rich. In general, rationing is based on the idea that there is a minimum standard of diet, both in amount and quantity, below which we cannot afford to allow anyone to fall. Are we, or are we not, going to apply the same principle to housing? Are we going to say that there is a certain minimum standard of accommodation, comfort and beauty below which we as a nation cannot afford to allow anyone to fall?

I personally hope so. This should be a part of the basic platform of security which every citizen is guaranteed by his country. Not only that, but without such provision we shall never get the increase in the birth-rate that we need. At the moment our population is not replacing itself and will shortly begin to decline. If we don't provide sufficient houses where parents can bring up a reasonable-sized family in comfort and without constant financial sacrifice, the decline is likely to be rapid and perhaps catastrophic.

To sum up one may say that in regard to housing, scientific research has opened the door to new possibilities. It is indeed essential; but by itself it is not enough. All it can do is to tell us how to get what we want. We must find out what we really want; and we must finally have the will to get it.

2

THE SCIENCE OF BUILDING

By Sir Edward Appleton, K.C.B., F.R.S., Secretary of the Department of Scientific and Industrial Research,
and
Sir George Burt

I'm not a practical builder, and you may therefore wonder why I should presume to talk to you about building houses. But I am a scientist, and it's what science can do in the building of homes that I'm going to talk about. I would, however, like to make it clear at the start that I believe that building is both a science and an art. We scientists can only formulate the scientific principles which must be observed in the building of a house so as to give its occupier shelter and comfort. But we leave plenty of latitude for the architect, while still obeying those principles, to express himself in its design and lay-out. So please don't think of the scientist as a man whose chief aim is to condemn you to houses which may be efficient, but which are all alike and simply dreadful to look at. That's not his object at all.

But there's really a third party interested in this question of houses besides the scientist and architect, and that is you, yourself. In a recent broadcast a number of people said what they wanted after the war. They wanted first, a job, and second, a decent home. I'm glad therefore to be able to tell you that the staff of the Building Research Station of the Department of Scientific and Industrial Research are working on many problems intimately connected with the building of houses for after the war.

They're dealing, for example, with such matters as heating, ventilation, plumbing, artificial lighting, day-lighting, noise reduction, and so on. In their work they are carrying out accurate experiments on bricks, on walls and rooms, and even on specially built houses. In this way they're gaining knowledge and finding ways of improving the quality and comfort of houses, without having to rely on guesswork.

Let me take an example to show you how a scientist tackles one of these problems. We've all suffered from the sort of house in which it's quite impossible to settle down to read in peace and quietness when there's a wireless on in the next room—or in which anyone having a bath late at night wakens up all the rest of the family. The problem is, of course, that of the transmission of sound and the scientist had to turn detective in tracking the various ways in which the sound vibrations travelled from one room to the next. Careful experiments showed that the sound does not all come straight through the dividing wall. It also travels along the floor and along the side walls. So if we want to stop this passage of sound we must go as far as we can to separate one room from the next. We must aim at designing our house so that it's not a box divided into rooms by partitions, but really a collection of separate boxes held together by a strong framework. There's no reason why the same principles shouldn't be applied to a lesser extent in any house or block of flats. That's the sort of way in which the discoveries of the scientist can help the architect.

Another problem that's received careful examination is that of artificial lighting. Science has been very busy in this field in close partnership with industry, and has now placed at your disposal a really remarkable series of improvements in lamps and fittings. But to take advantage of these, you've got to be a discriminating purchaser and resist the temptation to buy some other elaborate or ornate fitting which may catch your eye in the shop, but which is only half as efficient in lighting your room when you get it home.

Then, also, a great deal of scientific work has been done on ordinary domestic fires. The low efficiency of some of them is really disgraceful. This should be regarded not only as a drain on the owner's pocket but also as an unjustifiable waste of precious fuel. There's simply no need for us to have in our new homes fires that send most of their heat wastefully up the chimney or through the fire back to an outside wall. It may surprise you to learn that the room heating efficiency of many fires could, by scientific design, easily be doubled or even trebled.

I've only time to give you one more example of the work of our scientists and this time its a case of depressing, rather than raising, expectations. You've no doubt heard people saying that after the war we shall see houses built entirely of "plastics"—those new materials which are so light and so easily moulded—and which Sir Lawrence Bragg discusses in the next article in this book. But by making mechanical tests it has been possible to say whether such materials can be used to replace say steel girders or wooden joists in buildings. Unfortunately, the plastics have not got very high marks in this test. I cannot go into details, but I think it's safe to say that although plastics will, of course, find many uses inside the modern house it's fairly clear that it's unlikely that our houses will be built entirely of such materials.

Now let me hand you over to Sir George Burt, who is not only a practical builder, but also one with a real appreciation of what scientific research has done and can do to help him.

SIR GEORGE BURT, Director of a firm of Building Contractors

Since the days when our forefathers were content to live in a cave there is no doubt that we've advanced, but it is also true to say that up to comparatively few years ago the advance was made by rule of thumb methods. As general education and standards advanced, so, too, our requirements have risen. It is not very many years since the standard of one bathroom in the larger-sized houses, and certainly no bathroom at all in the smaller ones, was accepted as all that was necessary. It is perhaps twenty years since it began to be accepted that it might be desirable to have baths in every house; but now it must be very unusual for any house to be built without one. They have not yet all got running hot water, but we live in hopes.

The amenities that perhaps a few people hoped for in the old days—the luxuries of days gone by—are now looked upon by everyone as necessities—light, drainage, power, water, sewerage; but even so, there are still areas in the country where many thousands of people live in houses without piped water, and many more without main drainage, light or power supplies.

I think, perhaps, that you don't realise the extent to which Science can help in seeing that these things which might even now be called, and certainly up to a very few years ago were called, "refinements," can be given easily and without necessarily any appreciable expense. I suppose the most common source of annoyance to you is the noise of your neighbour's wireless, either in the flat above or in the next-door house. There is no reason for this; science to-day has information which can prevent it. It may be true to-day that this noise abatement might cost money, but there is no reason why it should always do so, and this is true of other sources of annoyance. Why do your walls run damp after a frost? Why, when there is even only a few degrees of frost, do your pipes freeze? All these annoyances can be easily avoided—and, if science is made to play its part properly, it needn't cost you any more money. What is really wanted is a better understanding between the authorities responsible for building houses and the architects and the builders; and, what I think is perhaps the most important of the lot, a firm insistence by you that you are going to have these things put right, and that you don't expect to pay more money for them. You must get the building industry in its very widest sense to realize that, and make them understand that you will see that you do get those things done as you want them.

It's up to you. Science is there waiting for you to insist on it being used.

Advances like this cannot be restricted to the building industry. Industries which have grown up since the war have got to come into it. You have got to use the knowledge acquired in aeroplane building and small ship construction and the hundred-and-one things—new things—which the war has shown us and taught us. So many of them can be adapted to help give you that better house which you want. Who would have thought that the fastest bombing aeroplane in the world—the Mosquito—would be made of plywood and balsa and glue, with nothing but its ailerons made of metal. Many of these materials which the war has developed can be brought to peacetime uses.

In house building we are inclined to be one-track minded; tradition is good but it can be overdone. Your brick-built house of to-day is stronger than it really need be for mere stability—it will certainly long outlast its usefulness as a habitation. Long before it is worn out it's likely to be completely out of date. While it may continue to give you shelter it won't give you those amenities that you will increasingly come to expect. There must be tens of thousands of houses in the country to-day which although they may not come within the Slum Clearance Order are quite definitely hopelessly out of date.

At the same time in striving after new things we are inclined to forget the virtues of the old. In the old days if you built a house more than one storey high you had to use mass—thick walls—to give it stability. With the improvement in materials and the craze for cheapness that mass was cut down; but in cutting that mass down you lost qualities which have never been put back in any other way—qualities of insulation from sound, insulation from cold and moisture. In the old-fashioned house you remember walking in from the outside on a bitterly cold day or a sweltering hot one and feeling a sense of relief that you have got into a more comfortable temperature; but the modern house doesn't give you that to-day. It can roast you in summer and freeze you in winter. Science will show you how this can be avoided, and how you can work to the old standards if science is allowed to play its part and proper use is made of new materials and new ideas.

To take another example—this time rather different. One can hardly fail to doubt that the war will enormously increase motor transport—more roads and better roads will be wanted. It seems to me that road building has progressed very slowly. We have certainly been spending more and more money on road-building in recent years but it's been spent on the surface. Your road specification normally lays down certain stipulations, but actually the only precaution taken in putting a road on a soil of doubtful or unknown strength at present takes the form of a thicker, heavier and, therefore, more expensive top, instead of really taking the trouble to find out what your soil will actually carry, and then building a road to suit it. The knowledge of science has not yet been properly applied. No architect or engineer would put a structure on a doubtful foundation, but there doesn't seem the slightest hesitation in doing so with a road, and spending more and more money in putting on a heavier and stronger top in a blind hope that it will last. This is certainly one branch where science has not yet been given a chance to prove its usefulness, and in road building there are many more.

We must hope that everyone concerned will come to realize the importance of what is known as the science of soil mechanics, and see that the money is spent where it ought to be spent—on the foundations—and not wasted by putting heavy, expensive tops on bad and unsound bottoms.

At the end of the war—immediately after the war—housing demands are going to be far greater than the industry as at present constituted can hope to cope with;

and it seems to me that there may be a serious danger of a lowering of standards rather than the raising of them. In my view there's no necessity for this. Let those responsible use the scientific knowledge which is at present available and you can have better houses than you had before the war—if not more quickly built, at least more quickly fit to live in; by which I mean they will take a shorter time to dry out and they won't necessarily be more expensive—and when I say "necessarily" I mean in relation to every other standard.

We've learned from the mistakes made after the last war, and there's no reason why these mistakes should be repeated. One bar to progress has been, perhaps, that you are too conservative in your requirements, and in most of your habits, Most of you are very wedded to open coal fires, but you forget, or perhaps don't know, that most of your heat is going up the chimney. With a little more flexibility in your requirements, and a few simple structural alterations in your house, this waste heat might be made to warm the whole house more thoroughly. If your children come home from school with homework to do, they want somewhere to do it quietly. In the winter to-day it would probably be too cold for them to use their bedrooms—but why should it be? We ought to encourage appliances that can help you not to let so much waste heat go up the chimney, and at the same time utilize that heat to better advantage where it is wanted. But to achieve this you'll have to be rather less set in your ways. Science can help in solving these problems, but your co-operation will be needed in getting the results used; and your interest in what is being done will ensure that proper use is being made of this knowledge, and that you are not being fobbed off with something which because it was good twenty years ago is good enough to-day. It's up to you to see that you get all the advantages which present-day science is able to give you.

In a broadcast such as this, I should be foolish if I entered into the realms of cost and finance. I'll only express the hope that we shall not endeavour to squeeze down costs to the pre-war slump level, extracted out of an industry by exploiting the position of its 20 per cent record of unemployment; it will indeed be a tragedy if we allow our standards to fall and let the fruits of scientific investigation over years rot on the ground. Let us set our standards high, let us realize that we must achieve these standards quickly and at a reasonable cost. I for one am convinced it can be done if we go the right way about it.

3

PLASTICS

By Sir Lawrence Bragg, F.R.S., Cavendish Professor of Physics in the University of Cambridge

What are plastics? We hear the word "plastic" very often nowadays. You will have noticed that many objects of daily use which were formerly made of natural substances are now being made of some kind of artificial preparation. Buttons are a good instance. When they were not made of metal they were generally made of bone, or of shell like the mother-of-pearl buttons affected by the Coster Pearly King. They are now made of an artificial bone-like substance, a plastic in fact, and in addition to the humble button of the trousers variety, one can get them in all sorts of shapes and sizes and brilliant colours. Toothbrush handles are of a tough transparent plastic instead of bone. Their bristles are of nylon—another plastic—instead of pigs' bristles, and they last much longer. Knife handles, mugs, the handles of umbrellas and the clasps of handbags, electrical fittings like plugs and switches, the case of your radio set, are made of plastics. The stiffeners in ladies' stays used to be made of whalebone. If there still are such things as stays, of which I am not sure, one would not now have to rifle the mouths of whales to get their vital parts. I remember my surprise when I first heard of a chemist friend with whom I was playing golf that the tees I was using were made from milk. By some sort of chemical juggling the plastic expert starts with such ordinary things as coal, limestone, salt, water and air and produces strong, light durable materials which can be used for a host of purposes.

The very fact that plastics are made of such ordinary materials, so unlike the final product, shows perhaps best of all what has been achieved. The chemist is copying nature. Grass feeds on water and air, and with the help of sunlight and a speck of mineral salts it grows its blades. The sheep eat the grass and turn it into wool. We have learnt how to carry out in the laboratory and factory what is done by living matter. Plastics fill a gap in the list of materials we need which used to be filled by such things as bone and horn and even wood.

To explain the main feature which is the basis of all these plastic materials, I must ask you to bear with a few words on the way in which chemical compounds are built up, particularly those compounds which are called organic because they play so large a part in living matter.

Here is an analogy to help us to see how chemical molecules are built by atoms. Most people are familiar with Meccano sets—it is rare nowadays to find a boy who has not built things with this delightful invention. There is a set of standard parts, such as different lengths of metal strip, angle pieces, plates with holes in them, and they are fastened together with nuts and bolts. With just a few standard parts one can build up many complicated structures such as bridges and cranes. Now atoms are like the standard parts in Meccano, while molecules are the structures built from them. The nuts and bolts which fasten the Meccano pieces together are the chemical bonds which fasten atom to atom in the molecule. The number of different kinds of atoms is not very large, some ninety or so in all, but, of course, by combining them in various ways an infinite range of chemical compounds can be formed. Now in organic compounds in particular, practically the whole structure is built of atoms of four kinds, carbon, oxygen, hydrogen and nitrogen. These are common atoms—coal is mostly carbon, water is hydrogen and oxygen, air contains oxygen and nitrogen. In some organic compounds other atoms play an essential part, just as in Meccano most of the structures can be built up of the ordinary pieces but we need certain special bits such as gear wheels to finish off our crane. Sulphur, phosphorus and chlorine are examples. The complex molecule called haemoglobin which takes the oxygen from the air and conveys it in the blood to all parts of our bodies, is a big structure of about ten thousand atoms. In this vast array, there are four atoms of iron which have an essential part in doing this work. Such special atoms, however, are very few. The main structure is always built up of carbon, oxygen, nitrogen and hydrogen. They are so useful because they have got convenient links, like the Meccano nuts and bolts. Carbon in particular has four available strong links for atom-connection, it is what we might call the fundamental standard bit of Meccano in all organic compounds.

Now the chemical molecule is a definite structure containing a certain number of atoms linked in a certain way. Water, for instance, is a collection of molecules each of which is two atoms of hydrogen linked to one of oxygen. A benzene molecule is six carbon atoms in a ring, each with an atom of hydrogen tacked on to it. Each kind of molecule is like one of the pictures in the Meccano book of instruction, some definite bit of structure which can be made from the parts. But now imagine a boy let loose with a whole set of identical units with their bolts, who idly set about joining bits all over the place, going on and on till he had a tangled mass on the floor, all of which was tied together in a random way. There is no end to the process, he could go on as long as he had spare parts. He would have done exactly what the chemist does when he makes a plastic.

How is a plastic made? In many cases the chemist starts with quite small molecules, so small that they slither easily over each other and make up a liquid. By some means these molecules are then made to link up with each other in all directions, so that they become a kind of tangle which is stiff and solid. Heat alone is often sufficient to do this. In making table knives, for instance, the steel blades are stuck into slits at the bottom of a nest of little boxes, which are moulds for the handles. These are filled with liquid and the whole is heated, when the liquid solidifies into an ivory-like solid gripping the spike on each blade. This sounds very simple, but of course, the secret lies in finding liquids whose molecules behave in this way. Perhaps I can best explain the kind of molecule which does the trick by my Meccano analogy. It must be a molecule which has connecting links to spare, but these connecting links must be, so to speak, hidden away inside it and not active till we are ready to use them. If this were not so, the molecules would be all sticking together at the start, whereas the whole object in plastic making is to start with something liquid or soft which can be poured or moulded, and then turn it into a hard solid. Suppose in Meccano we had a host of paired parts, each pair being fastened together by two bolts. They would all be separate and slide over each other quite easily (like a liquid). But by undoing one bolt per pair, we could use the spare bolt on pair A to fasten it to B, and then take the spare bolt on B to fasten it to C, and so on indefinitely, getting long strings. If this is done in a random way, the result would be a tangle of such strings so interwoven that it would be a solid mass. I will use another analogy because it is so important to get this main idea clear, which is at the bottom of all plastic-making. Imagine a number of couples dancing in a hall. Each couple is a molecule, joined together by a double bond, their arms. Since there are no links between one couple and another, the whole mass of dancers is fluid, and movement is possible. But now suppose one link in each couple be broken. Each lady and gentleman has one hand occupied with holding on to her or his partner, but the other is free and can clasp a similar free hand of a neighbouring couple. There will be such an intertwining of clasped hands that movement would become impossible. The fluid dance turns into a solid mass, composed of criss-crossing chains of people holding hands.

This is just what happens in one whole class of plastics.

Carbon atoms can be linked together in pairs by double bonds. By heating such molecules, and so knocking them about violently, one of these bonds can be opened, providing a spare link to which the molecule can be attached to a neighbour, and then to another and so on. Another trick to unite the molecules is this. A small bit is knocked off each of two molecules, leaving a link to spare, and this is used to bind them together. For instance, a hydrogen atom can be knocked off one, and an oxygen atom together with a hydrogen atom off another. The discarded atoms unite happily to form a molecule of water which goes off on its own, and the links so set free tie the two molecules together. If each molecule then has a hydrogen which can be knocked off at one end, and a hydroxyl (that's the name for oxygen and hydrogen) at the other, they are like the male and female screws at each end of a set of drain rods, and the process of uniting them into long strings can go on indefinitely.

This endless linking up of simple molecules is no new discovery. Nature invented it long ago. In plants, for instance, the chemical processes somehow produce a simple kind of sugar unit. These tack together by shedding off water in the way I have just described, and built up the long strings of cellulose. Some plastics make use of these long cellulose strings provided by nature. A chemical agent is used to loosen the strings from each other, when the cellulose mass becomes soft so that it can be moulded or squirted into threads through a die. The loosening chemical is then removed and the mass hardens in its new shape. A very old friend of this kind is celluloid, made of nitrated cellulose with the help of camphor. It was discovered in 1869. More recent ways of doing the loosening by chemical means have resulted in one class of artificial silks, and also in the cellophane which is so widely used nowadays.

The living matter in animals and plants is called protein. Protein molecules are exceedingly complex and their structure is still largely unknown, but we do know that they have a kind of backbone consisting of a long chain in which two carbon atoms alternate with a nitrogen atom. In living matter these chains are built regularly into the huge protein molecule, but a random tangle of them, no longer alive, is used by nature to build hair, silk, wool, and horn, nails and hoofs, all of which might be called natural plastics. The casein from milk is a protein, and can be used in the same way to make artificial plastic.

Most of my listeners must have made a plastic at some time themselves. You do it in fact every time you use an oil paint. Oils are composed of rather short chains of carbon atoms. When the oxygen of the air gets at certain kinds of oil like linseed oil, it builds cross links between these chains and turns the whole into a solid. We say "The paint is drying," but actually it is setting as a plastic.

Before I leave this question of structure I should say something about the difference between two main kinds of plastic. One kind of plastic becomes soft if it is warmed up, and hardens again on cooling. It can be dissolved by suitable chemicals and thrown out of solution again. The other kind, once it has set, has set for ever. It cannot be softened or dissolved. What causes the difference? It is a simple matter of the number of links. The first kind, the thermo-plastics, have two links per molecule. With two links, all one can do is to build up long chains. They may be so tangled up as to make the plastic hard, but they can still be separated or slide over each other when the whole is warmed up. The second kind, the thermo-setting plastics, have more than two links, say three or four. Probably you will see at once that not only can they join in strings, but also these strings can be linked sideways to each other in all sorts of ways. Such a mass cannot be softened or dissolved unless it is completely broken down. The cabinet of your wireless set is probably a plastic made from phenol with three links and formaldehyde as a cross-link; it is an example of a thermo-setting plastic.

Of what service are these plastics in everyday life? Think first of the properties of a plastic, its cleanliness, permanence, lightness and unbreakability. Then there is the ease with which many of them can be given a colour which is not on the surface but part of the material. Finally, since they start as liquids or powders and then harden, they can be made into any shape. It is a wonderful combination of properties. Instead of natural textiles like silk we can use artificial silk and rayon and nylon, which are cheaper and stronger. Natural textiles can be improved—for instance, crease-resisting cotton is made so by soaking with liquid materials which are then hardened into a plastic inside its fibres, making it springy. Transparent plastics are used for wind-screens and turrets of aeroplanes—lenses for cameras and spectacles are being made of plastic. One of the most interesting developments is that of using plywood with plastic glues. The wood is used as very thin sheets gummed together with special plastic cement. In this way one can build sheets of any rounded shape with no joints, as in the plastic aeroplane. Not only are the sheets very strong, but they are also very lasting and resistant to weather. This opens up all sorts of possibilities in the decorative side of housing and furniture. Artificial rubber is a plastic. Nearly every natural product—rubber, silk, guttapercha, bone, horn, can be replaced by a synthetic plastic. Are we emerging into a plastic age? We must be careful to keep a sense of proportion. Although plastics are made of cheap and plentiful raw materials, they require new and complicated machinery to handle them. Ordinary materials of construction, like the brick, are still very cheap and plentiful—we are not likely to go all plastic. Our architects and designers must be won over to apply their art—the exciting possibilities of a new material generally lead to atrocious bad taste in its employment at first. We may therefore expect to see plastics not so much displacing existing materials, but adding to them to an increasing extent in articles we use in everyday life.

4

CLOTHING AND FABRICS

By Professor J. B. Speakman, D.Sc., Professor of Textile
Industries in the University of Leeds

The fibres used in making clothing may be animal, vegetable
or mineral in origin. Silk and wool are animal fibres;
cotton and flax are vegetable fibres; and asbestos, which
is used in making fireproof clothing, is a mineral fibre.
Every one of these fibres has a complicated structure, but
the scientist wasn't encouraged to take an interest in textiles
until after the end of the first world war. It's an ill wind that
blows nobody any good, and it was because the country
couldn't wage a successful war without help from the scientist that our industrialists were led to believe that science
might be able to solve some of their peace-time problems
as well. Since 1918, scientists have been at work on each of
the main textile fibres and it is now fair to say that textile
technology is no longer a craft but an exact science. Apart
from textile *designers*, the training of those who will occupy
responsible posts in the industry must be essentially scientific
in character. The well-being of the whole industry depends
on the recognition of this truth.

Faced with the problems of the industry, the scientist
began to build up an exact knowledge of the composition,
structure and properties of the fibres concerned. At first,
progress was slow, because Nature is the greatest of chemists
and fibres are among the most complex of her products.
So the attack had to be a combined operation by the physicist and the chemist, and their joint work has given us a
very clear picture of the structure of most textile fibres, even
wool. Why even wool? Well, because it is only 20 years since
a distinguished scientist used to tell his students that the

structure of wool is so complex that it would never be discovered.

We now know that all fibres are made up of very long molecules, which are arranged more or less parallel to the length of the fibres, and are held together sideways with varying degrees of firmness. Where the sideways cohesion is small, as in the case of cotton, the fibres are easily dissolved and the solutions are used to make rayon. For a long time the chemist has been anxious to dissolve wool, too, because then he could make rayon from waste wool and the rags of the rag-and-bone merchant. If this could be done, it would revolutionize the shoddy trade, or, to give it its proper name, the Low Woollen industry.

Actually, the chemist has discovered two good ways of dissolving wool, but only after long and difficult work. Wool is extremely difficult to dissolve because in this case the long molecules are linked together by strong chemical bridges; the structure is something like a ladder, where the sides of the ladder are the long molecules and the rungs are the chemical bridges between them. But the structure is not rigid like that of a ladder, because the long molecules are folded down in concertina fashion. The consequences of this concertina ladder arrangement are very important.

When wool fibres are pulled, they stretch very easily because the coiled molecules unfold, or, if you like, because the concertina opens out. When the stretched fibres are released, they contract, because the molecules fold up again. If the fibres are damp, they return exactly to their original length, even after they've been pulled out to nearly twice their length. The damp wool is just like a piece of elastic, and this is the reason why fabrics made of wool do not crease easily during wear. If they do crease, the creases disappear when the garments are hung up for a short time in a wardrobe.

The crease-resistance of wool made chemists wonder whether other fibres, like cotton, which creases very easily, could be given the same property. The long molecules of cotton are not folded and when the fibres are stretched, or bent, or folded, the molecules slip over one another and the stretch, or bend, or fold, becomes more or less permanent. It was difficult for the chemist to make the cotton molecules fold up like those of wool, but he could bind them together so that they didn't slip over one another. He made simple chemicals react together inside the fibres to form a plastic —a plastic which you can look on as a kind of reinforcement of the weak cotton fibres. When the fibres of cotton fabric are permeated with plastics in this way, they show a crease-resistance similar to that of wool, and the discovery—it was made by Manchester chemists about sixteen years ago—has made cotton fabrics far more serviceable than ever before; and the process is even more effective with some of the rayons.

Unfortunately, the perfect elasticity of wool, which is so valuable in preventing creasing, is by no means an unmixed blessing. It causes one of the two kinds of shrinkage to which wool fabrics are liable. After newly-woven fabrics have been washed, they are stretched to standard width and dried in this state. If the fabrics were sold in this condition, they would naturally shrink as soon as they were laundered, or even on becoming damp in the rain, because the stretched fibres would contract. So before the fabric is sold, it is wound on to a perforated roller through which steam is blown. The steam escapes through the fabric and has the effect of annealing the fibres in their stretched state, so that they are afterwards unable to contract, even in water. The process is a very old one, but the exact reason why the fibres can be annealed in this way was discovered only a few years ago. I must give an outline of what happens in order to explain how the chemist was led to develop a number of improved types of wool.

When the wet fabric is stretched to standard width the folded molecules uncoil and the concertina opens out.

Steaming breaks down some of the rungs between the long molecules of the ladder-like arrangement and lets the fibres relax. When the rungs are broken, the fibre is seriously weakened, but fortunately for the user of wool textiles, further steaming causes new rungs to be built up again in new places while the fabric is in its stretched state. These new rungs prevent the fibres—and the fabric—contracting in water. As soon as the chemist had found out how the old rungs are broken, and how the new ones are built up again, he at once saw that both processes could be helped by means of simple chemicals, and more effective methods of preventing one kind of shrinkage were the result. Not only so, but he devised simple methods for breaking down the rungs and rebuilding new ones at ordinary temperatures, instead of in steam.

Research is full of surprises, and these new methods of fixing the size and shape of fabrics at low temperatures seem likely to simplify the permanent waving methods of the hairdressing trade. Chemically, wool and hair are very similar, and present-day permanent waving methods are very similar to the setting processes of the wool textile industry. The hair is wound on a curler, just as the fabric is wound on a perforated roller, and it is steamed in the curled condition. When the rungs between the long molecules of curled hairs are broken, the fibres relax and the curled state is afterwards made permanent by the formation of new rungs, just as in the case of the stretched fabric. Since the breakdown and rung-building processes can now be carried out at ordinary temperatures, who can doubt that it will soon be possible to give hair a permanent wave at ordinary temperatures, without the discomfort of present-day heat treatments?

In ways like these, the chemist was led to investigate the possibility of replacing the rungs in ordinary wool by others which would remove some of the weaknesses in wool which offend the user. Ordinary wool is very easily damaged by alkalis, and no one would dream of boiling wool materials in a solution of washing soda, say. In fact, no one but a member of a decontamination squad would dream of boiling wool garments in water. This is because one of the rungs in wool is very easily broken by alkalis and hot water. To overcome this weakness and make wool useful for many new purposes, the chemist has altered the rung and made it highly resistant to alkalis. So the chemist is making wool, as well as cotton, much more serviceable.

But this isn't the only importance of the work. These rungs in ordinary wool which are so easily broken by alkalis and hot water are very easily broken by the digestive juices of the moth grub. This allows the grub to use wool as food, and even if I can't eat my hat, the moth grub can. The new rungs which the chemist has been able to make in wool cause the grub to suffer from chronic indigestion, and some of the chemically modified wools are completely immune to attack by moths, besides being resistant to the alkalis they may meet with in laundering. So greatly improved wool will be available for general use after the war, and it is good to know that the methods for achieving these very desirable results were first evolved in this country, though they have since received a lot of attention in the United States of America.

Before leaving the subject of moth attack, I must mention that other methods of giving complete protection against moths had been developed by chemists in most countries of the world some years before the war. One such method was to dye wool with a colourless dye which is poisonous to the moth grub but not to human beings. In contact with such wool the grub must die either of starvation or of poisoning from the first bite taken. The treated wool has no smell and, in the words of a well-known advertisement, "only the moth can tell" that it is indigestible, and the reward of his discovery is death.

Wool has another peculiarity, which is sometimes useful

and sometimes a plain nuisance. Wool and hair are very similar, and if you have any hair left after five years of war you can see the effect I want to talk about by taking a hair from the head and rubbing it lengthways between finger and thumb. The hair travels out of the fingers because its surface is made up of flat cells which overlap one another like the slates on the roof of a house. Under a rubbing action, the surface acts like a ratchet and causes the fibre to move towards its root end, away from the tips of the protecting cells. When a wool fabric is rubbed in soap solution every fibre begins to creep in this way and the cloth shrinks until in the end it is a felt. In making fabrics like blankets, this property is extremely valuable because it allows the fabric to be consolidated in readiness for the next process, where the surface is raised with teazles to give a covering of loose fibre for warmth. But with socks and underwear the shrinkage is nothing but a nuisance. It can be prevented by treating the fabrics with a solution of chlorine in water. What happens is that the chlorine breaks down the very same rungs between the long molecules of wool as are attacked by alkalis and moths. When these rungs are broken down in the surface of the fibre, its skin swells in soap solution, and the ratchet—which is waiting to cause fibre movement—is made so soft and pliable that movement and shrinkage are impossible. As soon as we knew exactly how chlorine prevents felting, other ways of bringing about the same result were more or less obvious. To cut a long story short, there are now at least ten new methods of making wool unshrinkable, and the fact that they are all British is, I like to think, because the molecular structure of wool was first worked out in this country.

Even more striking examples of the practical value of scientific work on textile fibres can be found in the rayon industry. At first, the chemist's aim here was simply to make a cheap imitation of real silk, which has such remarkable strength and lustre. As all fibres consist of long molecules, the chemist chose as his raw material the long molecules provided by Nature in substances like cotton and wood pulp. The cotton is dissolved in various ways, one of the more important being discovered by Cross and Bevan many years ago, and the solution is squirted through tiny holes into a hardening medium. Here the long molecules are thrown out of solution as fine filaments, which are stretched to draw the molecules into line and to promote the rung formation which is needed for strength. The rayons produced in this and similar ways have found an established place in the textile industry, and the chemist is now busy trying to make fibres which resemble wool, at any rate in some of its properties. Wool is a protein, and the long molecules he uses in his efforts to find a substitute are obtained from milk, monkey nuts and soya beans. The fibres made from milk are now being used in the manufacture of felt hats because, curiously enough, mixtures of wool and milk-fibre felt much more quickly than wool alone, even though the milk fibre doesn't possess the ratchet-like surface structure of wool.

But the ambition of the chemist knows no bounds, and the latest types of synthetic fibre are made from long molecules which he himself makes from simple substances. You'll recognize an echo here of something Sir Lawrence Bragg has written in the previous chapter about Plastics. Some of the early products were very odd. I forget which bird it is that is supposed to fly backwards to see where it has been, but the chemist seems to have indulged in the same kind of aerobatics, because some of the first fibres just disappeared in soap solution, like some of the earlier rayons. A strongly developed sense of humour is needed to appreciate such materials, but they will at least be useful to the conjurer. All such difficulties have now been overcome and the American fibre nylon, which is made from benzene, is remarkable in being at least as strong as real silk. Its invention has spurred on the older rayon industries to further endeavour with the result that they, too, can now provide fibres which are considerably stronger than silk. Such materials have found important war-time uses, and it is now obvious to everyone that synthetic fibres must no longer be regarded as inferior imitations of natural fibres.

So far the chemist has not been able to make a good imitation of wool, but his attempts have begun to worry the wool-growing countries. Sheep-breeding will always be an important industry in Australia, South Africa and New Zealand, because of the demand for mutton, but wool provides a large part of their revenue, and it is important to consider what steps such countries can take to meet the competition of synthetic fibres. One safeguard is the more intensive prosecution of research on wool, which, as I have tried to show, is capable of correcting the defects of the fibre and widening its field of utility. A second lies in research on the by-products of the greasy fleece, particularly wool fat, which represents about 10 to 20 per cent of its weight and is commonly met with as lanoline. The chemistry of compounds related to wool fat is now well advanced, and the time has surely come for the chemist to be given an opportunity of making wool fat as valuable a product as the wool with which it is associated.

5

EXPLOSIVES

By PROFESSOR JOHN READ, F.R.S., Professor of Chemistry
in the University of St. Andrews

YEARS ago, as a boy in a Somerset village, I sometimes amused myself and impressed my friends by filling a jam-jar with water, turning it upside down in the village horse-pond, and poking the mud beneath it with a stick. Bubbles of marsh-gas, released from the bed of the pond, rose through the water and soon filled the jar. When I held a lighted match near the mouth of the inverted jar, the gas took fire and burnt quietly with an almost invisible flame. Occasionally, however, when a good deal of air had got into the jar, the mixture exploded with a very pleasing pop.

It's a long way from my village horse-pond to Hamburg and Berlin, and still further, you may think, from marsh-gas to "block-busters"; but the links in the chain are clearly traceable, so why not let us see how they run?

Marsh-gas, which arises from sodden and decaying vegetation, is the simplest of hundreds of thousands of organic compounds. Organic compounds are substances containing carbon. Marsh-gas, also known as methane, is a hydrocarbon, or compound containing carbon and hydrogen only. It is the same as the fire-damp of coal mines. Its molecule, or ultimate chemical particle, is written CH_4, because it is formed by the combination of one carbon atom, C, with four hydrogen atoms, H_4. These molecules are excessively minute. A hollow pin's-head would hold enough of them to provide several million each for every man, woman and child in the world.

When marsh-gas burns, it undergoes a chemical change known as oxidation. Through reaction with oxygen, which forms about one-fifth of the surrounding air, the carbon of the marsh-gas is burnt to carbon dioxide and the hydrogen is burnt to water vapour. So there are two oxidation processes going on together; the burning of carbon and the burning of hydrogen. We know from common experience that each of these processes liberates energy in the form of heat: consider a glowing brazier of charcoal on the one hand, and an oxy-hydrogen blowpipe on the other. These tell of the enormous stores of heat evolved in the burning of carbon in the brazier and of hydrogen in the blowpipe. So it's not surprising that marsh-gas, which contains carbon and hydrogen in combination, should burn with a hot flame; or that coal-gas, which is mainly a mixture of hydrogen and marsh-gas, should do the same.

When marsh-gas is burnt in a jar, or from a gas-jet, the heat is quickly dissipated through the surrounding air; moreover, the burning takes place quite slowly. The molecules of marsh-gas have to queue in the tube behind the jet, and wait for their ration of atmospheric oxygen until they quit the tube and get out into the air.

We reach a very different result when we mix marsh-gas in a closed space, with twice its volume of oxygen, and spark it. There's no queueing of molecules now. Each marsh-gas molecule has next to it the two oxygen molecules it needs. The reaction, started by the hot spark, spreads fiercely through the mixture. The equally sudden liberation of heat makes the mixed gaseous products so hot that in their efforts to expand they may shatter the confining vessel with a loud report. In common speech, the mixture explodes.

And now for a word of warning to any unskilled person who may think of trying experiments with explosives: it's the same as Mr. Punch's advice to those about to get married —Don't! With explosives, the consequences may be even more serious. The French scientist, Dulong, lost an eye and three fingers in explosives research—and, remember, *he* was a skilled and experienced chemist:

The many perils that environ
The man who meddles with a siren
Are naught beside the ones that he
Invites, who flirts with TNT.

An explosion is usually an exceedingly rapid oxidation, or burning. An explosive is a material capable of developing a sudden high pressure by the rapid formation of large volumes of gas. The explosive power of marsh-gas mixed with oxygen is relatively feeble, because the expansion is due entirely to the heat effect: except for the heat set free in the process, the volume of gas in this example would be the same before and after the burning. Enormously more powerful effects are produced in the explosion of suitable liquids or solids. A given space will accommodate a much greater weight of an explosive in the liquid or solid form than as a gas. This economy of packing in liquid and solid explosives is one of the leading factors in producing great pressure when the explosive suddenly gasifies. The second factor is the simultaneous liberation of vast stores of heat, leading to a further expansion.

When solid gunpowder explodes, it produces 500 times its own volume of gases, measured at the ordinary temperature; but the liberated heat causes a further eightfold expansion to 4,000 volumes. Nitro-glycerine is still more powerful: one volume of this oily liquid gives rise on exploding to 1,200 volumes of gas, expanding again about eightfold through the action of the generated heat to some 10,000 volumes. That is to say, a thimbleful of liquid nitro-glycerine is transformed in the twinkling of an eye into 60 pints of gas at a fierce heat exceeding 5,000 degrees Fahrenheit. On the same scale, a foot-rule would leap out to a length of two miles. Once started, nothing can stop or moderate this sudden burning and release of energy.

Why is nitro-glycerine so much more powerful than gunpowder? Briefly, because in gunpowder the fuel and the oxygen are done up in separate packets, or molecules; whereas in nitro-glycerine both the fuel and the necessary oxygen are packed together in the same molecule. In gunpowder, the fuel specks of carbon and sulphur lie side by side with the oxygen supply, contained in separate specks of nitre. In nitro-glycerine, the fuel atoms of carbon and hydrogen are arranged in the same molecule with a sufficient number of oxygen atoms for their complete burning. The mixing here is done inside each infinitesimal molecule: it is of the most intimate nature we can imagine. Most modern explosives are of this kind: nitro-glycerine, guncotton, cordite, tri-nitro-toluene, all contain the fuel atoms and the oxygen atoms arranged within the same molecule.

The molecules of such explosives are very delicately poised. The fuel atoms are temporarily held apart from the oxygen atoms by molecular policemen, consisting of atoms of nitrogen. Their lot is not a happy one; for they must always be on duty. The moment these pillars of molecular law and order relax their vigilance, there's a molecular dog-fight, virulent and contagious, and the countless legions of molecules collapse, with spectacular unanimity.

There are several ways of distracting the attention of these molecular policemen, so that explosion may occur. Sometimes heat does it, sometimes friction, sometimes concussion. Explosives are very temperamental. For instance, both tri-nitro-toluene and cordite will suffer the impact of a bullet without exploding; but mercury fulminate explodes when struck by a hammer, and nitrogen iodide is so very touchy that one would hesitate to sneeze near it, and a fly using a crystal of it as a landing-ground might no longer interest a spider. Again, cordite and tri-nitro-toluene burn without exploding when ignited in the open air, but mercury fulminate and lead azide explode with great violence when ignited under any conditions.

Many of the powerful modern explosives can only be roused to full explosion by means of detonation. In this process of detonation, discovered by Nobel in 1864, the explosion of a small charge of an initiatory explosive or detonant, such as mercury fulminate, sets off the main explosive lying near it. These detonants, which may be exploded by percussion or a spark, set up violent shock waves. So, when tri-nitro-toluene is fired by a suitable detonator, instead of burning quietly it undergoes an instantaneous collapse, caused by an explosive wave; this originates in the detonator, and moves through the TNT at a speed of more than four miles a second. This is what happens when a bomb or shell explodes.

Explosives such as gunpowder and cordite, which always burn comparatively slowly, without detonating, may be used as propellants. It's gunpowder that sends the shot after the rabbit and cordite that sends the bullet after the German. Other explosives, such as TNT, lyddite and guncotton, burn rapidly to detonation when confined. These are known as high explosives. They cannot be used as propellants, because they would detonate and shatter the weapon. High explosives are used for filling shells, bombs, torpedoes and mines, and also for demolition work.

A few miles from the fine horse-pond I mentioned just now, there is a grey old town called Ilchester. It stands at the junction of the Fosseway with the Roman road to Dorchester; but Ilchester is older even than the Roman roads. Here, in 1214, was born Roger Bacon, the earliest of the great scientists of England. It was in a Latin text, written in 1242, that Bacon first made known the composition of gunpowder, the oldest explosive. This Franciscan monk was perhaps the first to make gunpowder explode and to realize its power. According to a medieval legend, another monk, the mysterious Berthold Schwarz of the Black Forest in Germany, first used it as a propellant.

The introduction of modern explosives, beginning about 1850, was due largely to the Swedish chemical engineer, Nobel, who invented dynamite, blasting gelatine and ballistite, and founded the Nobel Prizes—including the Peace Prize.

Modern explosives are made chiefly from fats, cotton and coal, all of which are natural sources of energy, containing the fuel atoms, carbon and hydrogen. In the manufacture of explosives, the glycerine from fats, the cellulose of cotton, and the benzene and toluene of coal-tar are treated with nitric acid, under special conditions, in order to introduce the necessary oxygen and nitrogen atoms into the molecules. The nitric acid, formerly obtained from Chili saltpetre, is now prepared chiefly from atmospheric nitrogen. Indeed, to the Germans—because of British sea-power blocking the importation of Chilean nitrate—the manufacture of nitric acid from the air was essential before the war of 1914–1918 could be undertaken. So they made sure of it before committing themselves.

Through the intervention of plant life, fats, cellulose and coal also originate from the air, this time from gaseous carbon dioxide and water vapour. Nitric acid, fats, cellulose and coal—they all come ultimately from the air. Explosives are thus slowly woven from atmospheric gases, unto which, in the moment of explosion, they return—shedding suddenly their fabulous stores of strangely acquired energy, caught up mainly from solar radiation by the living plant.

How does all this that I've been talking about affect you? Economically and industrially, the manufacture of explosives is closely linked with the production of such familiar commodities as fats, glycerine, soap, cotton, coal, dyes, drugs, petroleum, and fertilizers. In the great chemical industries depending upon coal-tar, for example, explosives form one of many groups of fine chemicals, including dyes and drugs. These are so closely interlocked that a member of one group is often a by-product in the preparation of a member of another group. So, in the war of 1914–1918, Great Britain was sorely handicapped in producing explosives because of the lack of a strong organic chemical industry, and a corresponding dearth of skilled organic chemists. After the war, the position was safeguarded by a Dyestuffs Act, which prevented Germany from resuming her old practice of dominating the British fine-chemical market by underselling. It says little for the public appreciation of scientific problems in Great Britain that the renewal of this Act hung by a thread in 1937—a little more than two years before the start of a new war by Germany.

In the popular mind, explosives mean solely the propulsion of missiles, the bursting of bombs and shells, and destructive activities in general. Let us remember, however, that besides their destructive abuses in war, explosives have constructive uses of the highest value in peace. Many vital industrial and engineering operations would be impossible without their aid. In peace-time such civil activities as quarrying, mining, tunnelling, and the construction of roads and railways utilize explosives in hundreds of thousands of tons every year. Under proper control, explosives have an unrivalled capacity for doing useful work, as we may see in such wonderful constructions as the Simplon Tunnel and the Panama Canal. Economic, political, and even geographical considerations are clearly bound up with such achievements.

Think of the revolution accomplished in the excavation and removal of rock through the use of blasting explosives! Formerly this was done painfully by hand, with hammer and chisel, supplemented by "fire-setting," or splitting and flaking the rock by means of fire and cold water. Think again of the impossibility of mining coal for modern needs without the help of explosives! Consider, too, the incessant research providing ever safer explosives for this purpose.

It has sometimes been urged that we should abandon explosives and explosives research, because man has misapplied the discoveries. This position is untenable. Apart from the difficulty of securing international agreement in such a matter, there is an inherent urge in the human mind "to follow knowledge like a sinking star." It is no more possible to ban scientific research than to forbid exploration, mountaineering, or crossword puzzles. Remember also the inter-relations of explosives: coal-tar constituents are the common parents of TNT, lyddite, saccharin, synthetic indigo, salvarsan, and M. & B. 693. Even a controlled production of "key" chemicals—such as ammonia, nitric acid, and sulphuric acid—used in making explosives, is complicated by their position as "key" chemicals in numerous essential industries, including agriculture.

Scientists naturally deplore, even more than others, the perversion of their own discoveries and the debasement of their work and genius. Listen to the eighteenth-century Dutch scientist, Boerhaave, the most famous physician, the foremost chemist, and the most erudite scholar of his day. The art of war, he observed in 1732, has turned entirely upon the one chemical invention of gunpowder. "God grant," he added, "that mortal men may not be so ingenious at their own art, as to pervert a profitable science any longer to such horrible·uses."

When man discovered fire, he took into his keeping an instrument with unbounded possibilities for good—or evil. Fire, says the old proverb, is a good servant but a bad master. But we have not banned the use of fire because a cigarette-end thrown carelessly into a rickyard may destroy a whole harvest. Explosives are a refined form of fire.

Let us end where we began, at that instructive horse-pond. Beside it stood the village smithy. "Your fire's out!" I said one day to my friend the smith. He stroked the long bellows-handle caressingly, and a glow soon appeared in the embers. "Out, is ur?" said the smith, "Why, zonny, there's vire enough in he vur to burn down all London!" Whereupon he thrust an unfinished horseshoe into the midst of the glow.

"The fault, dear Brutus, is not in our stars, but in ourselves."

This being so, may we not look forward to an enlightened age, in which the discoveries of science will be used entirely for the benefit of mankind?

It is worth our while to consider what practical measures we can take to realize this ideal.

Berthold Schwarz in his laboratory (*see page 45*)
(*From a copper engraving by R. Custos,* 1643)

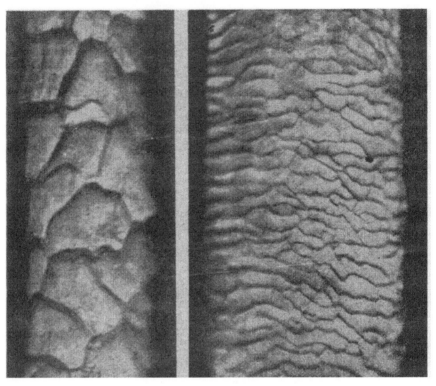

Surface structure of (*left*) wool, magnification 800, and (*right*) human hair near the root, magnification 500 (*see page 38*)

6

SOUNDING THE EARTH'S CRUST

By Dr. A. O. Rankine, O.B.E., F.R.S., Chief Physicist
of the Anglo-Iranian Oil Co.

When gold was discovered in Australia, sizeable lumps of it were frequently found exposed on the earth's surface. They say even now in Bendigo that after rain you can pick up in the neighbourhood of a pound's worth of gold particles —if you're lucky and patient enough to spend a day at it. In Persia—or Iran as it's now called—I've seen the natives filling their vessels from a little pool fed by a bubbling spring. But it wasn't water they drew; it was oil—white oil—fit without refining to put in their lamps. Everyone knows, too, that some of our coal is won from surface workings where it lies exposed to view and is dug up almost as one quarries chalk or sand.

But such easy finding is rare. The treasures of the earth are mostly buried beneath the surface, often at great depth. Some mines are nearly a mile deep, and oil has recently been obtained from holes bored close on three miles into the earth's crust. The question is, how can we tell where they are. What other means besides seeing are there for locating these hidden riches, so that they may be extracted from the earth with a minimum of effort? Every unsuccessful boring—a "dry" hole as we call it in oil prospecting—means waste of materials, and often many months of fruitless labour. That is where the geologist and the geophysicist come in.

What I have to say has nothing to do with divining as commonly understood. The methods we geophysicists employ depend upon well-established physical principles.

Generally speaking, the closer you are to a thing the more

likely you are to find it. This is true of search by geophysical instruments; when they're near what is being looked for they can give reliable indications of the objects sought. When they're not near enough they can't help at all. In fact, geophysical instruments have a restricted range. It would be quite impracticable, for example, to use them to locate oil-bearing rocks without any previous guidance as to the areas within which those rocks are likely to be. We must first get help in choosing areas worth examining geophysically, and we get this guidance from the geologist who is, in fact, the precursor of the geophysicist. Geological studies of the composition of rocks and the way in which they lie in the earth's crust enable the geologist to select in many cases comparatively small regions under which the mineral sought probably lies. In his task he relies largely on the fact that the sequence of rock beds or strata which constitute the terrestrial shell do not lie everywhere horizontally. They have in years long past been bent or folded into mountains and valleys, and thereafter worn away where prominent so that rocks elsewhere deeply buried are exposed to view and scrutiny. The nature of these outcropping beds, particularly the manner in which they slope, helps the geologist to predict how they extend underground. But as the distance from the outcrop increases, the more vague becomes the geologist's picture of the underground structure. It's here that the task of the geophysicist begins. He has to survey the neighbourhood with instruments so as to define the structure more precisely and thereby fix the points where drilling or digging should start to reach the buried objective. This collaboration between the geologist and geophysicist is truly a combined operation, and the more likely to succeed the firmer it is.

How then does the geophysicist play his part? Fundamentally, he has got to rely on the thing which he seeks being in some way *different* from the rock beds which cover it. Suppose that the search is for oil deep down. In this case it proves to be impracticable to make use of the special physical properties of the oil itself; but geological experience indicates that certain rock beds, such as limestone, are likely to be impregnated with oil if they are humped upwards, so as to form, as it were, an underground hill. The problem thus becomes one of locating the summit of this buried limestone hump. There are two physical methods of doing this by observation carried out on the earth's surface.

The basis of the first method is the force of gravity. We are all familiar with this force; we've learnt to attribute the pull downwards that we feel when we support a weight to the attraction the earth exerts upon it. But do we realize that every portion of the earth, even the smallest, contributes its part to this force, to an extent depending upon how much there is of it and how far it is away? Dense rocks also exercise at the same distance a greater attraction than those less dense. Now limestone is usually denser than the geological strata overlying it; hence volume for volume it exerts greater gravitational force. Thus if we were to carry a weight over a buried limestone anticline—one of these humps, as I've called them—it would get heavier as we approached the summit of the hump and lose weight again after passing over it.

I'm not suggesting that this is the actual method of procedure. The variations of force in question are far too small to be felt in this way. Our sensation of pull is, indeed, particularly crude, and some much more delicate means of observation must be found. Even our much more powerful sense of vision sometimes requires instrumental aid, as when we use telescopes or microscopes. For accurate perception of force, here with greater necessity, we have to employ special forms of apparatus, called gravity meters. These gravity meters depend generally upon the twisting, bending, or alteration of length of fine metal wire or delicate springs suitably loaded, under the influence of gravitational changes. They have the requisite sensitiveness, but are nevertheless robustly made; they have to be, so as to stand up to the rough usage inevitable in field operations. The best of them will measure changes in gravity of less than one part in ten million—about equivalent to one more drop of water in your already full bath. I have now vividly in mind two remarkable gravity meters which I saw being operated a few years ago in the State of Kuwait on the Persian Gulf. In the course of only seven months the whole of Kuwait's five thousand square miles—about the size of Yorkshire—was surveyed in great detail. And the gravity picture thus obtained helped a lot in fixing the location of highly productive oil wells sunk subsequently in that country.

It mustn't be supposed that the gravitational method is limited to the mapping of *limestone* underground. Under suitable conditions it can be applied, whenever the object sought is *different* in density—whether less or more—from the things that surround it. The method has indeed achieved its greatest success in the United States, where, under the plains around the Gulf of Mexico, numerous domes of rock salt, with which oil is associated, have been located.

But difficulties arise when the surface of the region to be surveyed is rough and hilly. The hills themselves exert gravitational forces which may be large enough to mask the feebler forces due to the underground irregularities. In these circumstances one has to turn to an alternative method—the second I shall describe, which has, in fact, been much more widely used. It's called the seismic method, or, if you like, the "earth-shaking" method.

To illustrate it, think again of a potentially oil-bearing limestone anticline like those occurring in Iran. The difference as between limestone and its overlying beds upon which this "earth-shaking" method depends is not now that of density, but what, for want of a better colloquial term, I shall call hardness. The limestone is harder than the strata above it; consequently, in spite of its somewhat greater density, mechanical shocks travel through it faster than they do through the overburden. Moreover, this same difference of hardness causes shocks coming from above to be partially reflected at the surface of the limestone. Either or both these phenomena may be made use of by those practising seismic prospecting. They have to be equipped with very delicate seismometers, which are also robust and portable. These instruments record earth tremors and the times at which they reach the points of observation. To create the tremors, charges of dynamite are exploded in the earth's surface at suitable places. As regards both the measurements of time and what is deduced therefrom, seismic prospecting resembles closely the better known subject of seismology—the study of the internal structure of the earth as a whole by means of the tremors originating in natural earthquakes. Only our earthquakes are very much in miniature, besides being started just when and where we want them by exploding anything from a few pounds to several tons of gelignite.

The particular procedure most commonly employed, and the easiest to describe, is to observe what are in effect echoes; only it is the seismometers, and not our ears, that "hear" them and measure the time elapsing between the instant of the explosion and the arrival of the echo. The velocity of the tremors in the overlying beds being known from suitable measurements, the depths of our limestone at any chosen points can be found by simple calculation, and its shape thus disclosed. It's by no means always so straightforward as this. Often there are complications due to multiple reflections from other beds, and great skill is needed in the interpretations of the seismograms—seismograms, that's what the records produced by the seismometers are called. The method fails not infrequently, but it succeeds, too, often enough for it to be regarded by progressive oil companies in the United States as an essential preliminary to boring for oil. The same is true of the extensive operations by oil companies in this country as well as Iran, where seismic prospecting of a somewhat different type, based upon the *refraction* instead of the reflection of the seismic waves, has been practised with considerable success.

Geophysical prospecting is a comparatively new application of science; it began to be practised on a notable scale less than twenty-five years ago. Although now in great vogue, its surveys are done chiefly in regions sparsely populated; consequently, its direct effects on everyday life are not much noticed. Where the surveys do encroach on human habitations they do make their presence felt—quite literally if the method used is the seismic one. The exploding charges cause some slight damage to crops and occasionally to livestock; but the compensation paid by the prospecting party is usually more than adequate. Some claims, of course, have to be rejected, as when the injury is said to be the reluctance of hens to lay, owing to fear of our little earthquakes. On the positive side, although it is as yet no great matter, it may be recorded that the visit of a prospecting party nearly always provides employment for local workmen, who are needed to assist the geophysicists in their field operations.

But it is not this direct effect upon human life that deserves emphasis. We should consider rather the question of ultimate utility, and ask whether this sounding of the earth's crust has contributed to our well-being, through the produce of the earth which it helps to make available to mankind.

I have confined my remarks in this talk mainly to the search for oil, and said little about other minerals, partly because I know more about oil, but also for the reason that I wanted to answer in the affirmative the question just posed. Without geophysical prospecting we should either be suffering now from a world shortage of oil, quite apart from war-time restrictions, or at least have to face scarcity before long. The dangers of this situation have been realized by oil producers generally, and money measured in millions of pounds has already been spent on geophysical work, particularly in the United States, the source at present of most of the world's oil supply. Moreover, extensive and intensive geophysical campaigns have been initiated in many other countries hitherto untapped, in regions geologically suitable for such examination. It's reasonably certain that much more oil lies hidden in the earth than has been taken out of it—a trifle of less than two cubic miles —and geophysics is sure to go on helping to make it available.

And if crude oil as it comes from the wells is in copious and increasing supply, so also will be the useful things derived from it—petrol, synthetic rubber and plastics, to mention only a few of many.

As to minerals other than oil, such as the ores of iron, copper and lead, there has been so far relatively little demand for the help of physics in searching for them. Appropriate methods have, none the less, been developed quietly and form, as it were, weapons in reserve. They depend upon the marked magnetic and electrical properties of these ores. If these so-called base, but most useful, metals should threaten to become scarce, the geophysicists, newly-equipped, may be called again in strength to the prospecting front, and may be expected to give an equally good account of themselves.

We've been thinking of finding minerals—a matter not easy to discuss without some technical knowledge. What to do with them when found is of more general interest. Perhaps you may care to consider this, remembering how very different in political development are the countries in which the minerals are found, and try to answer the question— What organization of distribution and use would conduce most to the general benefit of mankind?

7

OUR WEATHER

By Sir Nelson Johnson, K.C.B.,
Director of the Meteorological Office

"Here is the Air Ministry's weather forecast for to-morrow. There will be occasional rain in most districts, but also bright intervals. Thunder will occur locally. It will continue rather warm."

That was the last forecast issued by the B.B.C before the outbreak of war. In those care-free days, most of you thought of the weather in terms of its effect upon your garden, or the Test Match, or in the case of air travellers that flight you were making to Paris.

If you ask the Air Ministry for a weather forecast now you won't get one—indeed, they will tell you practically nothing about the weather.

We all know, of course, that weather forecasts are of importance for war purposes—particularly for the R.A.F., but I think it may be interesting to see in just what ways the weather affects flying operations. So let us follow the day's work of some of the meteorological officers in Bomber Command.

First thing in the morning the Commander-in-Chief at Headquarters, Bomber Command, sends for his Senior Meteorological Officer to learn the general weather situation and to select the night's target accordingly. The selection may be a provisional one, because at that time there may be some uncertainty about the exact weather to be expected eighteen hours later at a place possibly six hundred miles away.

By the afternoon further weather reports have come in, and make it possible to be more precise. The Senior Meteoro-

logical Officer at Headquarters now gets into touch with
the Senior Meteorological Officers at all the groups in
Bomber Command which are to take part in the operation,
and they proceed to hold a "telephone conference." In
this way the opinions of all these experts are pooled, and the
Commander-in-Chief and the various Group Commanders
can be presented with an "agreed" forecast. And on the
strength of this forecast the operation is either confirmed or
cancelled.

What factors do the meteorological officers have to take
into account when making their forecast? The weather on
the route to the target must be favourable—in particular
there must be no undue risk of severe "icing" of the aircraft.
On reaching the target area, the pilots must be able to see
their objective clearly—it must not be hidden either by a
layer of cloud or a blanket of fog or haze lying on the
ground. Last, but by no means least, when the aircraft reach
home again, the pilots must be able to find their aerodromes
still free from low cloud and fog so that, exhausted themselves
and possibly with a damaged aircraft, they can land with the
minimum of difficulty.

If the operation is "on," the meteorological staff at each
of the groups must then work out their own details—the
wind and the weather which their own aircraft will meet on
the way to the target, and the weather changes which are
likely to take place at their own base aerodromes by the time
the aircraft are due back.

Shortly before take-off, the pilots and navigators are given
the latest and most accurate advice possible. This briefing,
as it is called, is carried out by the meteorological officers
at the actual bomber stations. They explain the weather
situation in detail so that the air crews can act with a full
understanding of the weather changes they are going to
meet. Questions are asked and answered to clear up any
doubtful points. Now and again the briefing officer may be
able to say: "I think you will be home early; it looks as

though you will come in for a tail wind on the return trip."

But the meteorologist's task isn't finished when the aircraft have been despatched. He must see them landed safely back at their bases again. He must continue to watch the weather minutely throughout the night to see that it is going according to plan. If, by chance, early morning fog begins to form at some aerodromes earlier than expected, he has to make rapid decisions as to which aerodromes will remain "open" and for how long, so that aircraft whose aerodromes have gone "out", as we call it, may be directed to safety.

And when you remember that a bomber operation may involve nearly a thousand aircraft with some 5,000 men inside them, you will, I think, agree that the meteorological officer carries a big load of responsibility.

We chose for our example a bomber operation, but the same type of procedure is followed for all flying operations—whether it be the anti-submarine patrols of Coastal Command, or the daylight sweeps and night "intruder" operations of Fighter Command, or the trans-Atlantic flights of Transport Command, Indeed, it is safe to say that no flight is undertaken without due regard to the weather. And if a flight has to be made in spite of the weather, knowledge of how it is going to change may be of tremendous importance—both to the success of the operation and the safety of the air crew.

By what ways and means does the meteorological officer make his forecast for these purposes?

To make a forecast it is necessary to draw a weather map similar to those which were published in some of the daily newspapers before the war. But the forecaster's working chart is, of course, very much larger—about as big as a fully opened newspaper—and it covers a very wide area—most of Europe, the Atlantic Ocean and possibly North America too. Before the war there was no difficulty at all in covering the chart with weather reports because, by international arrangement every country broadcast its own reports for the benefit of every other nation. Numerous ships sent reports from the Atlantic and these were of special importance. The reason for the importance of these ships' reports is that, broadly speaking, most weather systems travel from west to east, so that to-morrow's weather in England is frequently decided by the type of weather out over the Atlantic to-day.

But with the outbreak of war the peace-time arrangements went by the board. The countries at war, and some neutrals, too, stopped broadcasting their reports, and wireless silence was immediately imposed on all ships at sea. As a result the forecasters found themselves with huge blank spaces on their charts, and it became necessary to devise new methods of obtaining information for making reliable forecasts.

In passing, it is worth noticing that the fact that weather travels from west to east makes it particularly important to prevent the Germans from knowing the weather condition in the British Isles, since it would give them an invaluable guide to the weather they could expect. That is why we put such a strict ban on our weather information. But to return.

The first and most obvious way of getting extra information is to arrange for R.A.F. aircraft when engaged upon operations to make and bring back weather reports. Bombers attacking Berlin or any other targets on the Continent do this, and bring back reports of the weather they have encountered, which are of the greatest value in forecasting the conditions to be expected over Germany the following day. In the same way, Coastal Command aircraft hunting submarines out in the Atlantic, and Transport Command machines flying over from North America, tell us what depressions and "fronts" are advancing upon the British Isles from the west. The zeal shown by the R.A.F. air crews in making these observations, when they already have a thousand and one other things to attend to, is beyond all praise.

But there are certain regions which are not visited sufficiently regularly by ordinary aircraft, and to cover these, special weather patrols are made. Specially trained pilots and observers go out every day—whatever the weather—in aircraft fitted with meteorological instruments. They fly long distances into remote areas stretching from the Arctic to the Azores and bring back most valuable weather reports. Any U-boat or Ju. 88's which they encounter are taken in their stride as part of the day's work.

To help you understand the next two points I am going to mention, let me remind you that the atmosphere is not just a shallow layer close to the ground—it extends upwards to great heights. And so we cannot obtain a complete picture of what is taking place if we have observations only of the conditions near the ground. Indeed, it has been clear for some time past that we shall never get to understand the weather properly until we have sufficient reports from the upper regions. It is in this direction that progress is to be looked for.

The most obvious way of finding out the pressure and temperature and humidity of the upper air is by going up in an aeroplane and measuring these things with the proper instruments. The R.A.F. have been doing this regularly for several years now, and they bring back as well reports on the clouds which they have met on the way up and down. All this information is of the greatest importance to the forecaster in working out the type of weather to be expected for to-night's or to-morrow's operation.

But there is a more modern method of doing the same thing more quickly and more cheaply. In this method we send up a free balloon about six feet high carrying a special instrument called a radio-sonde. This instrument measures the pressure and temperature and humidity of the air as the balloon rises, and then automatically sends out radio signals of the readings it has made. All the observer has to do is to sit beside a radio receiver in his office and receive *immediate* reports of the weather conditions all the way up to ten or twelve miles high. And since the balloon is carried by the wind as it rises, the observer—a different man in this case—can also measure the speed and direction of the wind all the way up if he "follows" the balloon with a wireless direction finder. And when, in due course, the balloon bursts, a parachute opens and the radio-sonde floats down to earth again. Each instrument carries a label telling the finder how to return it to the Air Ministry and offering him a reward of five shillings for his trouble. After slight adjustment the radio-sonde can then be used again to play its part in winning the war.

The Germans are also using instruments of this kind but the wording on their labels is rather different. Instead of offering the finder a reward, it threatens him with dire penalties if he fails to comply with the orders it gives him. The Hun runs true to type, even in little things!

So far we have been thinking about the weather mainly as it affects the R.A.F. But the other fighting services need weather information too. Every anti-aircraft gunner must know what winds and other conditions his shells are going to encounter, and he must make the proper allowances for them if he hopes to hit his target. The same is true of the other kinds of guns the army uses.

Sometimes the army try to locate the position of an enemy gun by what is called "sound-ranging." The method is similar to measuring the distance of a thunderstorm by noting the time between the lightning flash and the thunder. In this case you calculate the distance of the enemy gun by measuring the time between the flash and the bang. But to fix the gun position accurately, you must make the proper corrections for wind and other weather conditions.

The weather also plays an important part in the use of smoke screens. Indeed, it may decide whether a smoke screen can, or cannot be produced to cover a particular military operation.

The Navy, too, has its own special weather problems which are dealt with by its own meteorological organization.

Let us now forget about the war and look ahead to see what part weather knowledge will play when we return to peace.

Does anyone doubt the importance of civil aviation in the post-war era? Will not the bombers and reconnaissance aircraft which now fly such long distances over the continents and oceans give place to giant air liners flying over greater distances? Such methods of conveying passengers, mail and freight have come to stay, but their operation will not be possible without the weather service.

At the outbreak of the present war short distance civil aviation within the British Isles was just beginning to get into its stride. As soon as men and machines become available it will go ahead again.

Those of you who travel by these means may rest assured that your pilots and navigators will be supplied with the best weather information that experience and modern developments can provide.

There are many other ways in which weather affects us all—directly or indirectly. Climatic statistics are needed for town planning, and rainfall data are required by the water supply undertakings. The sites for civil aerodromes cannot be selected without knowing the frequency of low cloud and fog, whilst the lay-out of the runways must be planned in the light of the local prevailing wind directions.

The knowledge of future weather, and particularly spells of settled weather, is of great value to the farmer in deciding when to sow, reap or carry. Gale warnings protect our shipping. Forecasts can also be made of when the Thames is liable to overflow its banks so that steps may be taken by the authorities to prevent danger to human life. By anticipating snow and what is called "glazed frost," action can be taken in advance to keep the railways and roads open when otherwise serious dislocation of traffic would be bound to occur. The large electricity undertakings can be warned when thunderstorms are expected and precautions taken to protect their systems from the effects of lightning. These are some of the uses to which weather forecasts are already put.

Many business undertakings are beginning to realize that the economy and efficiency of their concerns can be increased by taking account of the weather elements, and we may feel sure that the return to peace will discover new directions in which weather knowledge can be placed at the service of the community.

8

THE HOUSEWIFE AND THE FISHERIES

By Michael Graham, Fishery Scientist, author of
The Fish Gate and *Soil and Sense*

For several years now, I have been anxious about the power of the housewife's shopping basket, which has increased so much with modern development of transport. There was a time when food mainly came from close to the house; for example, cows were kept in London itself. Alternatively, the housewife of those days bought fully preserved food, like salt beef.

But nowadays, in peace-time, a housewife can buy the second-best from any part of the world—and only a little stale at that. This modern development has made life precarious for the producers of fresh food; for example, the shepherds of the Downs, whose large sheep gave joints that were not as convenient for small households as those from the grazing type of sheep common in New Zealand.

Fish provides another example. To the housewife, in peace-time, fish is a small part of the menu—hardly noticed one way or the other; and in war-time it is so badly wanted that prices rise to quite foolish heights. But fish to a fishery scientist is mainly a problem of population—populations of fish—to be solved by some of the simpler mathematics.

I can make this clear by summarizing the history of modern trawling: at first, and for nearly a hundred years, fishing produced more fish from near waters by exerting more and more fishing power: more vessels, and larger and faster ones. Finally, in the North Sea, we have been taking every year about two-thirds of the average stock (young fish, of course, grow up to replace those caught).

Two-thirds of the stock every year is a large proportion,

and many people do not know that fishing operations dig so deep into this natural resource. It is therefore worth looking at the evidence for my statement.

Firstly, as to the quantity caught, it should be explained that there are very good statistics of the weight of the catch for many years past—since 1906. Collecting and tabulating them has been a very big task; but there is no doubt of its success. Thanks to conscientious and critical service, by devoted civil servants, the statistics are right to within a few per cent.

Naturally, it is more difficult to estimate the weight of fish in the sea; and if this can be done within 20 or 30 per cent, it is something of a triumph. For one stock of fish, the plaice of the North Sea, we have two independent methods.

One way is to count the eggs. The plaice gather to spawn in the southern North Sea just after Christmas each year, and their eggs float up to the surface, 250,000 from each female plaice. By sampling the area regularly during the season, with a silk net that filters a known volume of water, it is possible to estimate the total number of eggs. This can be used to give the total number of spawners, and, knowing the proportion of mature to immature, the total number of plaice of fishable size. The answer comes to 300,000,000 —in a good year.

The other way of estimating the number of plaice in the sea is to mark the fish and let them go again, and then find what proportion of the marked fish are returned in the catch. About 17,000 plaice were marked between 1900 and 1914, and about 5,000 marks were returned to fishery officers. But, of course, many marks were shed before the fish were caught, and others were kept by the fishermen, or lost. It is possible to estimate these losses reasonably well, and make adjustments that I need not enter into here. The result of this estimate was that the average stock was 210,000,000 fishable plaice. Taking into consideration that this is an average figure over many years, whereas the other was a figure for a good year, the two answers are in good enough agreement.

So the stock is known, and the catch is known, and the ratio of two-thirds, found in one method, is confirmed by the other.

Another kind of evidence, on the large proportion of the stock caught by trawling, comes from war experience. During the years 1914–1918 there was very little fishing in the North Sea, and, when we came to examine what happened as a result, we found that the weight caught per unit of fishing effort in 1919 was about twice that of 1913, taking comparable areas, and efforts. In this war, too, the catches of the few vessels left fishing round the edges of the North Sea have greatly increased; and, in addition, some Danish fishermen, captured on the Dogger Bank in 1943, reported to the Press that they could fill their vessels in half the time that they needed in peace-time. The effect of stopping fishing has been phenomenal in both periods.

Turning now to the commercial aspect—the effect of this severe trawling on the profits of the industry—we have clear and unassailable evidence from the report of the Sea Fish Commission of 1936 that the profit was driven out of the business. From the very careful analysis of accounts it was shown that Near Waters trawling was literally unprofitable, although many men stayed on in it in hopes of better times. The better times never came.

The explanation of this chronic lack of profit has been proved to be a simple one—that the fish were not allowed time to grow to any size. Now the war has allowed them to grow again. A great deal of research has gone to prove that moderated fishing would always allow them to grow to a reasonable size—when they would be heavy enough on the average to give the fishermen a fair living. The size would also be better for trade, and for the housewives. To sum up the lesson of fifty years' research work: until there is moderated trawling there will not be a good regular supply of fresh haddocks, cod, plaice, and hake. This is a serious matter in many respects. The fish are needed, especially by old people, by invalids, and by children; and the country has another interest as well, namely a steady fishing industry and a flourishing population of fishermen.

So far I have been dealing with the overfished stocks of trawled fish in Near Waters, but there are other fisheries whose state at present is essentially different, and it is in these that the power of the market—that is, the housewife's shopping basket—is seen most clearly.

Thus, turning to the herring fishery, there are millions and millions of herring, and catching them at the rate of a thousand million a year, as we did once, did not disastrously reduce the stock, so far as could be told by science; although it did, we think, make large herring somewhat scarcer. Herrings are of many races or tribes, and some herrings can be caught, in one or another locality round the British Isles, at any time of the year. But the greatest concentration in the world—of the best quality for inland trade or export —takes place in October and November only thirty to fifty miles away from the coast of Norfolk. This concentration is a remarkable natural resource, the wealth of which is borne in on anyone who has taken part in the fishery.

One day stands out vividly in my memory. The men started to haul the nets at four o'clock in the morning. They had looked at the first net or two and estimated that the catch of all the nets would amount to 100,000 herrings; in their own words, they had had a "look on" at the first net or two, in the darkness, seeing the flash of the herrings silvery sides in the lamplight; and had decided that there was a "hundred cran shimmer." So the word had gone to all hands to haul the nets. But at daybreak, when only a small part of the nets were hauled, the fish struck again; and by ten o'clock in the morning, we, and all our neighbours round us, knew that we had a tremendous catch. The fishermen in other craft judged by the time it was taking our ship to crawl slowly up the line of nets as we gathered them in. They also saw the great flurry of gulls round us, and about a score of gannets working our fish. The gannets circled 100 feet up in the air; chose their fish from one of the many beating away from the net below the surface; and then dived, squawking at the gulls to drive them out of the way, as they came down in the dive with their wings tense-shouldered for the shock of the water. Below the ship, too, the hunt was on. We could see the gladiator whales turning and swerving below us as they seized the dying herrings.

For us, excitement was suppressed by toil. The net was hauled and shaken, hour after hour, to the limits of human fatigue. Herring scales seemed to fly like sawdust; herrings were in the scuppers, on the engine casing, under our feet everywhere as we trod.

At noon our nets would come no more. The herrings had died in them, unable to work their gills, and the weight of dead fish had taken the nets down to foul some obstruction on the bed of the sea. So we had to cut; and another craft, not encumbered by any catch of her own, took over the task of lifting our abandoned nets, and their fish, starting from the other end of the line of nets.

Our crew worked for twenty-four hours, with only half an hour's break for tea, before they finished landing that catch. But it was the best of the season; and against it must be set many nights when we fished and caught nothing.

Herring, they say, at half a crown a pound, would be esteemed above salmon; and there is doubtless some truth in that. Cheapness, very naturally, can obscure the recognition of quality. Quality in food cannot easily be defined by scientific analysis, but fish is at last beginning to be recognized as "first class protein," and a herring is among the best of fishes. It is, however, much neglected. If everyone ate one herring or kipper once a week during the first half of the year, and thrice a week from Midsummer to Christmas, which by good natural arrangement is the season when eggs are scarce and dear, there would be no problem in the herring fishery. The price would have to be about 1½d. per fish. There would be a surplus in October and November, but our foreign trade could easily handle that.

Instead, our herring fishermen worked, for many years in the Dreary Thirties, for less than a pound a week. About half of them gave up altogether. Doubtless, herrings were rather smelly to cook in small households; doubtless, there are too many bones for hurried eating; and the herring does not come in a pretty package with a coloured wrapper. When, for other reasons, our export trade in herrings shrank disastrously, it is a pity that these other, petty, difficulties in the home trade prevented it saving for us the men and the ships that the Admiralty needed so badly in this war.

Another large fishery, which was not exhausted in pre-war fishing, was that for cod in the Arctic. There were undoubtedly thousands of millions of them, inhabiting many northern seas, but especially those round Iceland, Norway, Spitzbergen, and over to Nova Zembla. Those grounds are a long way away, and, unfortunately, by the time the cod reached England they were stale. For my part, I would not pay 4d. per lb. for them.

But cod was just cod, and money for food was tight, and because these stale cod, in their vast quantities, could be sold at the quay for half a crown a stone, really fresh cod from the North Sea and Faroes, had to fall in price, too.

Refrigeration was tried in the Arctic ships, but when this, decently fresh, cod went into the trade, it was just cod, and the cost of refrigeration was not recouped. It would have been necessary for shoppers to have known of this fresher cod, and to have been willing to pay a little more for it.

I write of fisheries because of their intrinsic interest, but, as my studies have progressed, I have found several things, "laws" we should call them in the scientific world, that seem to me to apply generally. One of them is the power of the shopping basket, to which the study of the Arctic cod fisheries has brought me back again. It is my idea that a very large proportion of the money that is spent by individual people is spent by women shopping. Consequently, it may not be an exaggeration to say that the greatest power in the world is the hand that fills the shopping basket. Every week the men hand that power over to the women, in the form of housekeeping money. If that much is true, or even somewhat true, we have a power here that can make or mar any plans for reconstruction; and we therefore have to beg women to be discriminating in their housewifery, and critical, and industrious—as some women have been in every generation.

The women appear to be the real masters, at whose bidding men mine the earth, and till the land, and fish the seas.

If it is true that the earth's natural resources can only be rightly used if women direct their shopping to that end; then I wonder how the world's needs can be made known, and thoroughly understood, and established, so that some at least of the shoppers may know what they are about.

I cannot give the answer to that; and I cannot decently press my own view, which is that science can help here.

The late Professor Karl Pearson wrote a book called *The Grammar of Science*, of which I have always liked the introductory chapter. "We must carefully guard ourselves," he wrote, "against supposing that the scientific frame of mind is a peculiarity of the professional scientist." This frame of mind includes "the insight into method, and the habit of dispassionate investigation," which "give the mind an invaluable power" of dealing with facts, as the occasion arises.

But he warns us against making too great a claim. "I am only praising the scientific habit of mind," which, he writes, will enable a scientist to judge in other fields, according as he has classified and appreciated his facts, and been guided by them and not by personal feeling and bias in his judgments.

Other contributors to this series have recounted the many services of science in technical aids and devices for better living. It seems to me that there is something to add. To my mind, science, properly used, could render society a supreme service, if it could engender a scientific attitude to the common problems of life.

9

SAVING LIFE AT SEA

By Dr. Albert Parker, Director of Fuel Research

In the year 1912, two years before the last Great War, the wreck of British ships at sea caused the loss of more than 2,300 lives. 1912—that was the year which included the heart-rending loss of the *Titanic*, when 1,500 passengers and crew went down. Since that time, in the short space of thirty years, science has made discoveries which have greatly-reduced the hazards of shipwreck in the vast ocean expanses. I need only remind you of the developments in radio and other ways of signalling, whereby those in distress can indicate their position to ships over distances of many miles. Then we have the modern aeroplane, which travels at such great speed, that it can rapidly search enormous areas. In consequence, the risks at sea, in normal times of peace, are now very small compared with what they were thirty or forty years ago, when rescue depended largely on the chance of being sighted by some ship passing nearby.

But it isn't always realized what science has done, and is doing, in improving the design of life-boats, and in lessening the discomfort of the occupants until help arrives.

What are the primary material needs of man? They are air to breathe, water to drink, food to eat, and warmth, clothing and shelter to protect him from the elements. Of these, air and water are of the first importance. It's well known that a man can live for several weeks without food, but he cannot live for more than a few days without water.

We in Great Britain, who have never been desperately short of water, cannot realize what real thirst means. But you ask the desert soldier who has been lost from his unit!

You ask the seaman who has been torpedoed and adrift for weeks! Some of those intrepid sailors have been shipwrecked more than once; they know what privation means.

During these war years, the problems of saving life at sea have received great attention in several quarters. As a result, important improvements have been made in life-saving devices. How best to provide drinking water has been specially studied by the Department of Scientific and Industrial Research, in co-operation with the Ministry of War Transport, the Medical Research Council, and industry. It is because it has been my job to lead a team of scientists in finding how best to make drinking water from sea-water, that I have been asked to talk to you to-night.

In tackling the problem of getting drinking water for life-boats, we must first know how much water each man needs a day and for how many days. In a temperate climate like that in this country, the average quantity of water taken in one form or another by an adult is roughly two-and-a-half pints a day. It is known, however, that much smaller quantities will maintain life for many weeks. If we set the high standard of say one pint a day for four weeks, 140 gallons would be required for a life-boat for forty people. It's not at present practicable to allow space in a crowded life-boat for 140 gallons of stored water, with all the other vital equipment.

Before the war, such a life-boat was provided with ten gallons of stored drinking water; but in peace-time, when ships follow recognized sea lanes, help usually arrives within a few hours or days. In war-time, help may be longer delayed, and it is now the practice to supply nearly thirty gallons of stored water, equivalent to nearly a quarter of a pint a day for each person for four weeks. This improvement has been made by skilful re-arrangement of the inside of the boat, and the stowage—an intricate problem to which much thought has been given. In wet weather the supply can be supplemented by collecting rain water. A fabric rain-catcher has been designed by the Ministry of War Transport. When rolled up for packing, the rain catcher is little more than a foot long and is only a few inches wide. When extended and tied down, it catches rain over an area of about six square yards. With heavy rain, twenty-five gallons can be collected in a few hours. But there is not always rain, and when there is none, something more is needed.

You may well ask why men should be short of drinking water, when they are surrounded by billions of gallons of sea-water? Unfortunately, sea-water isn't fit for drinking because it contains so much dissolved salt. It is, in fact, dangerous to drink sea-water, unless you also take a large proportion of other water containing little or no salt.

What is the amount of salt in sea-water and what is its nature? Over many years, there has been continuous scientific investigation of the waters of the open sea. In this work, which has been organized internationally, Admiralty and other British scientists have played a great part.

Sea-water contains a mixture of salts and it has been shown that four fifths of the mixture is the same as common salt used for cooking. Broadly speaking, the amount of salt in sea-water is between 3 per cent and 4 per cent. This amount is about four times as great as the salt in the human system and mainly explains the harmful effect of drinking only sea-water. To most people, water containing only one-tenth of the amount of salt in sea-water would taste horribly salty.

What is needed in emergency at sea, is some easy method of removing the salt from sea-water to give drinking water. There are several ways of doing this, but they are mostly difficult to operate. The most obvious way is distillation, that is to boil the sea-water and condense the steam. It is also known that when sea-water is cooled to freezing-point, the ice which first separates contains very little salt, because most of the salt remains dissolved in the water not yet frozen. But there are other ways. By adding certain chemicals, the salts can be separated as solids which can be removed by filtering the water; this method is complicated and uncertain unless very carefully controlled. Nearly ten years ago, it was found by British chemists that certain synthetic resins can take the salts out of sea-water. These resins belong to the class of substances known as plastics, which are described by Sir Lawrence Bragg in Chapter 3. By this method, the sea-water flows through granules of one kind of resin, and then through granules of another kind. The action of these resins is similar to that of the material in ordinary household water softeners, but they do a lot more. Unfortunately, the volume of drinking water obtained in this way from sea-water is not much greater than the volume occupied by the resins themselves, unless the resins are treated frequently with acid and alkali. There are other methods of removing salt, but they are very involved.

From recent work, it has been concluded that the two most promising ways are distillation, and a combination of chemical treatment with a method similar to that using resins. After numerous experiments, distillation has been selected as the best for life-boats.

It's not so easy as it may seem to design a small still for life-boats, as several important conditions must be met. The space occupied by the still and fuel must be much less than that of the distilled water obtained on operating the still for say a few hours a day for a fortnight; otherwise it is better to carry stored water. Simplicity of operation is essential and there must be no complicated mechanism likely to fail in emergency. Further, the equipment must be so designed that large numbers can be made quickly by mass-production methods, with very little skilled labour.

In the first place, several stills were designed to produce from one pint to four pints of drinking water an hour. These stills were made by highly-skilled technicians in the workshop. At the same time, experiments were made in co-operation with the oil industry and others to select a fuel and a burner to heat the stills. In the next stage, the equipment was used to distill sea-water from the North Sea. It was important that natural sea-water should be used in the experiments because it is more liable to froth and boil over than tap-water in which salt has been dissolved. Tests were then made in the workshop and in the open air under conditions similar to those in a boat at sea. To imitate the tossing of a boat the stills were operated while swinging in a pendulum, and to imitate a strong wind, a current of air from a fan was blown over the equipment. Later, tests were made by several groups of seamen using the stills in life-boats on the sea. Several improvements resulted. During this time the scientists worked in co-operation with firms who might make the stills and burners in large numbers. This is the stage at which the precision model of the research technician is converted to the mass-produced but efficient article.

Two types of still have been chosen for production and a third is under consideration; one type was designed by the Department of Scientific and Industrial Research, and the others were due to individual inventors. Each still produces about four pints of drinking water an hour; and the volume of water obtained when the still is operated for several hours a day for a fortnight is much greater than the volume of the still and fuel. One type of still is heated by an oil burner, and the others are heated by burning briquetted coal, wood, or other solid fuel.

Let us imagine that we are in a life-boat and wish to start up the still heated by oil. This still is in the form of a vertical cylinder with a central flue containing the burner. It is first clamped in position in the boat. Sea-water is poured into a reservoir at the top; this reservoir is like an unspillable inkwell, so that sea-water is not lost when the boat tosses. A turn of a screw opens the oil feed from a tank surrounding the still, to the burner, which is then lit. All that is then necessary is to add sea-water at intervals. After a short

SAVING LIFE AT SEA   77

time drinking water flows steadily from a pipe near the base, and continues to flow at four pints an hour so long as the burner is lit.

It is intended that stored drinking water shall continue to be carried in life-boats, and that the fabric rain-catcher and still shall be additional equipment. There should then be no real shortage of drinking water.

I've only told you about one or two of the developments made by scientists to meet the special conditions of war. There have been many other improvements. Though the requirements in times of peace are not so stringent as in war, some of these developments will find their uses also in saving life at sea, when the war is over.

H. S. HUMPHREYS, Chief Engineer Superintendent of the British Tanker Co. Ltd.

When an oil tanker is hit by torpedo or bomb, there is a risk that she may catch fire. If this did happen, oil might escape from the tanks causing the sea around the ship to be enveloped in flames, which would tend to spread with the wind, to leeward, over the surface of the sea.

Of course, everybody has always been concerned to protect our undaunted tanker-men from such hazards and the latest development is a new type steel life-boat designed by the Oil Industry—with the co-operation of the Ministry of War Transport.

A model of the boat was tank-tested at the National Physical Laboratory to obtain the best form, with optimum stability for seakindliness and to ensure the best propelling efficiency. A prototype boat was then built by a boat-building firm on the Clyde.

The first essential requirement is to get the life-boat away from the ship's side and clear of flames quickly and to protect the occupants from fire.

The prototype boat was designed to meet these essentials and to give good sailing qualities and to prevent, so far as practicable, undue exposure and exhaustion of the crew.

It was decided to adopt the open cockpit type of boat 28 feet in length, with a raised steel deck forward and aft.

The buoyancy tanks are built into the boat, forming a double shell.

A deep coaming extends all round the cockpit, which is protected by a fireproof sliding canopy.

Water sprayers, worked by two hand pumps, provide a constant spray of water over the whole external surface of the boat above the waterline.

The new boat can be propelled alternatively by Diesel engine, or electric drive, or Fleming hand-gear.

The hand-gear is operated by eight hand levers linked to the propeller shaft through a gear-box. The electric drive is operated by motor-car batteries.

The sails consist of jib, lug and mizzen.

The life-boat has a capacity for 33 persons, and, when fully manned and equipped, weighs about seven tons.

She is fitted with quick releasing gear and lowered into the water by gravity davits and flexible steel wire falls.

In the fully-loaded condition the ten horse-power Diesel engine produced, in the prototype boat, a mean speed of five and a half knots, that is about six and a third miles an hour. The speed with the hand-gear was four knots which represents about four and two thirds miles an hour.

Sailing trials and tests of the seating arrangements and stability entirely satisfied the Ministry of War Transport surveyors.

The most important test was that which required the boat to be subjected to intense fire and smoke for a period of four minutes; four minutes was laid down for the test because it was estimated that in that time the boat could be propelled, either by power or hand-gear, a distance of at least a quarter of a mile against the wind, guided by a wind indicator; and a quarter of a mile should take the boat beyond the limit of any oil which might be burning on the sea.

A large water tank, in which the boat was placed, was used for the fire tests. The surface of the water was heavily covered with oil and ignited.

During the final fire test the boat was occupied by personnel who had been concerned with its development. This test was extended to five minutes, during which time the flames reached heights of over forty feet, the boat being lost to view in smoke and flames.

The quenching and cooling effect of the water sprays, on which the safety of the crew and boat depends, was very noticeable. The occupants showed no signs of distress after their ordeal and intimated that conditions within the boat never became unpleasant; neither was the sea-worthiness of the boat nor the efficiency of the fire protection affected.

The Ministry of War Transport have now placed initial orders for 500 of these boats, together with the corresponding sets of gravity davits and equipment and have authorized conversion of existing boats to incorporate the fire-fighting provisions of the new boat, which represent an advance upon the arrangements previously provided.

10

SCIENCE AND SHIP DESIGN

By J. L. KENT, Superintendent of the William Froude Laboratory of the National Physical Laboratory

IT is not necessary for me to tell you why we need a first-class Royal Navy, because we've known for many generations past that our very existence as a nation depends upon the efficiency of our warships and the men who man them. But do we need a large and efficient merchant navy? Surely, the various unaccustomed things we do, and do without in this war, in order to save shipping-space, must have taught the most obstinate landsman amongst us our dependence upon ships, ships, and still more ships. A large merchant navy is vital to our existence and the Royal Navy is the policeman who sees that the ships of our mercantile marine can pursue unmolested their lawful business upon all the seven seas.

If the merchant navy is to exist in the face of the fierce competition which has attacked it in the past and may again, it must have efficiently designed ships, and the seamen must be efficient at their jobs. To secure such efficiency, science can be used in a multitude of ways in ship design.

The State enforces rules which are meant to ensure the ship's safety at sea, and certain others known as the tonnage laws. These latter are supposed to be a measure of the space available in the vessel for carrying cargo, and it is on this cargo capacity measure, that port, dock and canal dues are levied as a kind of income tax. These laws very greatly affect ship design, and they should have a strong scientific basis. Too often, however, in the past they have been based upon tradition and rule of thumb.

To design a merchant ship successfully, the naval architect

Showing the Alfred Yarrow Tank with the carriage spanning the waterway, and models

Top—New fire-resistant steel life-boat for tankers, showing water sprays being tested before the fire test

Bottom—The life-boat hidden in smoke and flames during the fire tests

uses every branch of science—the mechanical sciences in efficient engine and boiler design and also in hull design where the aim is to combine minimum hull weight with maximum strength and so squeeze a little more cargo-carrying space into the ship to earn more money per voyage. The physical sciences for navigation instruments such as echo sounding apparatus and wireless direction finders, as well as refrigeration or air conditioning of the ship's holds, which in recent years has made possible the sea transport of meat and soft fruits like bananas. The chemical sciences are used in the means taken to reduce corrosion and fouling of the ship's bottom, so increasing the life of the ship and preventing a serious drop in her speed. Even the medical sciences are studied in the lay-out of hospital ships and the sick bays of warships and liners. Mathematics is, of course, greatly used for stability, strength and resistance calculations, some of which are extremely complicated. The science of economics plays its part in the study of trade-route conditions and this has a great influence on the ship dimensions and design. So you see every branch of science must be used if the road to the perfect ship of the future is not to be strewn with costly failures.

Naval architects must have a good working knowledge of all these sciences if they are to succeed, and as may be expected they are, as a class, famous for their humility, for of all the professions, theirs is the one which daily emphasizes the depth of ignorance of the wisest men of the way in which Nature's laws work, and as Solomon put it, the way of the ship upon the sea is still beyond man's understanding.

The purposes for which ships are built are so many and so varied that each class presents its own particular problems for solution and a brief glance at a few of them will, I think, show how science is used to solve these.

Warships may have to carry large heavy guns at high speeds over long distances, work in compact fleets and be capable of offensive action day or night in all weathers whilst protecting themselves from hostile action of all kinds—bombs, torpedoes, mines and gunfire. To do this the variety of precision instruments of highly scientific design used in such craft is staggering. Submarines and air-craft carriers became practicable ships only after much scientific research on their designs, and such things as the sweeps used by minesweepers to clear the fairways of mines, were only successfully developed after much work by scientists. The war has seen the introduction of many strange looking vessels for amphibious operations, and the design of these assault craft presented many a knotty problem for the scientist to solve. In the merchant navy science ensures the passenger's comfort by giving him an even temperature in his cabin, which must also be free from objectionable noises and odours. These cabins are furnished to minister to the passenger's slightest need, satisfy his aesthetic taste, and all with an economy of space quite phenomenal. On the other hand, the horse-drawn canal barge was the subject of much full-scale scientific research in England over 110 years ago—experiments which have since become a classic in the profession, and even now science is re-shaping our barges. In recent years the resistance of the humble barge has been reduced by over 40 per cent without increase in cost or sacrifice of carrying capacity.

The defenders of the America Cup had the hulls and sails of their yachts scientifically designed and tested in a ship model experiment tank, and the result fully justified the use made of science in their design. Even the racing eight has been the subject of scientific design in order to secure low resistance, light weight and great strength.

To explain how modern science can be harnessed to ship design, let me describe one comparatively small, though highly important part of the work of designing a merchant vessel.

This is finding the most efficient shape of hull for the job required of the vessel, with due regard to cost, safety, capacity, practicability of construction, dimensions, wharfage and port facilities upon the route on which she runs, and a few other things. The aim must be to secure low resistance to lessen fuel consumption, so the science of hydrodynamics is evoked, and experiments with models bridges the gaps in our scientific knowledge. These experiments are done in model testing tanks, which are like huge swimming baths as wide as a road and about 200 yards long. They may be 10 to 16 feet deep, too, although arrangements are made to make the water shallow and to build up in miniature such waterways as the Suez Canal, or the mouths of some of the great rivers when problems connected with steering in confined waters have to be solved.

A large steel bridge, which may be anything from 5 to 45 tons in weight spans the tank and can be driven at high speeds along it, upon levelled rails fixed to the tank walls. This carriage is used to tow the models and carry the experimenter with the measuring apparatus. The model hull itself is usually made of wax, which is cast in a china clay mould and shaped in a specially designed machine. This machine cuts the correct contours in the wax casting while the operator traces them one by one upon a drawing of the ship design. These wax models, by the way, are not small affairs. They vary in length from 16 to 20 feet and weigh anything from three-quarters of a ton to a ton-and-a-half, and in special cases may reach 4 to 5 tons when fully loaded. I have known as many as three grown men carried in one of these models during special work. When finished, the model is towed through the water at various speeds, its resistance measured and the wave patterns it creates in the water are filmed and studied. This enables the naval architect to see whether the ship will get the desired speed economically and if not, his experience and training suggest changes in hull form, which are rapidly made on the model and tested. The propeller is designed on scientific principles and a model screw made and fitted to the hull. Further experiments are carried out with the model propelling itself at various speeds down the tank and from the data so gathered, the engine power, screw revolutions and efficiency are obtained. In the last fifteen years such tests have resulted in a 20 per cent reduction of the fuel bill of the modern cargo ship and has literally saved hundreds of thousands of tons of fuel. These experiments may be followed by such tests as steering to prove the rudder, or rolling to examine the efficiency of the proposed bilge keels. Or waves may be created in the tank by a special machine and rough water experiments carried out, when the pitching and heaving of the model are automatically recorded, the aim being to reduce the unpleasant erratic ship motions in storms and so save fuel and the passengers' appetites. These large experiment tanks are also used in war-time to perfect many ideas concerning things put into the sea to annoy enemy shipping, and to protect our own merchant and royal navies.

In addition to these large experiment tanks water tunnels are also used for the scientific study of propeller action under a heavy thrust—as in destroyers, say, where many thousands of horse-power are absorbed by each screw. These tunnels are very like the now famous wind tunnels used in the study of aeroplanes in flight, with the difference that water is pumped through the tunnel instead of air.

Yet another adjunct of the experimental tank is the steering pond in which large-scale models turn circles under their own power and record the efficiency of the proposed rudder design together with the power required to operate the steering gear.

When economic conditions rapidly change, as during slumps or booms in world trade, cargo ships hitherto very efficient, suddenly become uneconomical because their dimensions and the cargo capacity of their holds are unsuited to the new conditions. Then ship surgery is the only cure and an operation is performed on the hull. The ship is cut in two, and whole sections of it removed or new holds are built in as science dictates. Where the ship is to be cut, and the exact amount to be added or removed from her hull, is decided by careful scientific experiments in the ship model tank. Not so long ago a whole class of low power cargo ships trading to the Orient were shortened and some high power liners lengthened, so as to remain paying propositions in spite of drastic changes in world trade, and science once again proved its use to the business world. Experiments on ship models are followed by tests in the actual ship on her acceptance trials and maiden voyage. Then the scientists endure all the discomforts of sea and weather (and often internal uneasiness, too, as I know from experience) to secure facts which will shed light on obscure points of design. I recall one such case in the *Berengaria*. Her bridge was over 100 feet above the water, yet spray continually drenched the bridge deck. By studying the movement of snowflakes during a snow flurry this was shown to be due to an unpleasantly shaped bow wave far below, and was demonstrated by holding a length of rope over this wave with the vessel doing 23 knots. When let go the rope soared skywards up the tiers of decks and landed upon the bridge. This wave was entirely due to the ship lines and because of the expense nothing much could be done to alter it. But in subsequent vessels care was taken to avoid the creation of such waves. On another voyage for an hour during a violent storm in the Caribbean sea I was suspended at the end of a rope over the stern of a large tanker (which was performing dizzy and erratic gyrations at the time), in order to secure a photograph of the action of the rudder in a seaway, and this information gave facts which were used in future rudder design.

Now is the expenditure of all this effort, expense and time on the scientific design of a tramp steamer say, really necessary and a matter of national importance? By taking all this care, the cost of transport by ship of goods from abroad can be cheapened. With intensive scientifically directed research, goods hitherto not generally obtainable in this country could be transported cheaply and without damage from the countries where they are made or grown. The war has shown us that if we were reduced to the bare essentials to support life, civilization would take a long stride backwards. It is not too much to say that progress in civilization depends mainly upon an abundant stream of the ornamentals of life such as art in all its forms (painting, music, literature, and so on) including those unessentials of life which please the five senses, as for example, tropical fruits and flowers, scents and such things as tea, sugar and spices, rubber and so on which can only be obtained abroad. These can be made available to all to raise our standard of living, if we possess a cheap, plentiful and efficient sea transport, for science teaches that the cost per ton of cargo by air transport will always be many times that of water transport, if both services are economically independent and run without subsidy.

The ordinary man-in-the-street and especially his housewife, will benefit directly from a large fleet of scientifically designed and operated merchant ships. Cheap transport leads to cheaper goods in greater variety, which permits the housewife greater choice in the exercise of her aesthetic instincts and desires, when selecting her purchases, and so raises her family's standard of living.

For many centuries the design of ships was an art, and success depended upon the personal skill of the naval architect and the craftsman who built those ships. With the march of civilization this art is fast becoming a science and to enable the naval architect to design better and better ships, science must be increasingly used. This can only be done successfully if research is carried on unceasingly.

II

THE TUNNEL BUILDERS

By G. L. Groves, B.Sc., M.Inst.C.E.

The work of the tunnel builder is not only *at* your service, it is *in* your service, day and night, year in, year out. For civil engineering works are the bed-rock of our national life. And tunnels, together with roads and railways, bridges, docks and such like, make up an equipment—a sort of national tool-kit—without which our way of living would be completely upset.

Supposing, for instance, that, by some malevolent agency, all the tunnels in this country were put out of action, what sort of plight should we be in? To take the most obvious thing first, long-distance traffic on the main line railways would come to a standstill because the railway systems would be, for the most part, cut up into short, isolated lengths; and all London's tube services would cease to run. That would be bad enough for those of us who have much travelling to do. But far more serious would be the interruption of supplies —food, coal, munitions, and many other things that are more than ever vital to us nowadays. That, however, is by no means the complete picture. You who live in several of our larger towns and cities would lose most, if not all, of your water supplies, because they are brought to you from distant reservoirs by aqueducts which, in parts of their lengths are tunnels. (Big tunnels they are, too; a lorry could be driven through some of them if they were empty of water.) Another calamity would be the stopping up of sewage systems in scores of towns up and down the country; in most large centres of population parts, at least, of these systems will have been constructed by tunnelling. (I wonder if Londoners realize that there are upwards of 400 miles of *main* sewers alone in the vast drainage network of their city.) Again, there would be failure of electric supplies from those great power stations—and there are quite a few—in which the condensing water essential to the operation of modern turbine plant is circulated through tunnels.

Well, that is a dismal list of disasters. I could add to it, but you can see clearly enough how many necessary services, which we take pretty much for granted, owe something to the skill of the tunnel builder.

It was in the early part of last century that tunnelling began to develop scientifically as a branch of civil engineering practice. Two things happened then which are landmarks in tunnelling history—the invention of the tunnel shield and the construction of the Thames Tunnel.

I'll come to tunnel shields in a minute. Before doing so, let me tell you briefly of the fearful difficulties which had to be faced by the first engineer to link the two sides of London's river by tunnel. It is the first chapter of a story which is still being written—the story of big tunnels under wide rivers. Some important chapters are likely to be added to that story in the years to come.

In 1835, Marc Isambard Brunel started to build a tunnel under the Thames between Wapping on the north bank and Rotherhithe on the south. It was an ambitious undertaking for those times, for not only had nothing of the kind been attempted before, but the tunnel was to be of great size. It was to have a double roadway in one rectangular brick-work structure measuring nearly 40 feet in width and over 20 feet in height. The boldness of the project created a sensation, not only in this country, but abroad. The Duke of Wellington, referring to this, said, "I speak from my own knowledge when I state that there is no subject with reference to which the interest of foreign nations is more excited than the tunnel under the Thames; they look upon it as the greatest work of art ever contemplated."

But Brunel found himself in difficulties almost from the

start. The clay through which he expected to drive the tunnel turned out to be mingled with seams and pockets of water-bearing gravel, and with silt of a particularly unwholesome nature—unwholesome because the Thames was then London's main drain. This was a grievous set-back for Brunel and his work resolved itself into one long fight against the threat of the river to break in. Often the river did break in, twice so violently that it swept the workmen off their feet and flooded the part of the tunnel already completed. The second time this happened six were drowned and Brunel's son—later to become engineer to the Great Western Railway—narrowly escaped with his life. The workmen fell sick; they came out on strike; fire-damp added another anxiety; money ran short as the time spent on the work grew long. In 1829 the works were closed down for nearly seven years. When they were re-opened, with improved equipment, matters went a little more smoothly, although Brunel's untiring devotion to his task never lessened. Before the tunnel had been started he had moved from Chelsea to Blackfriars in order to be near his work; now, for the next four years, he had samples of ground from the working face of the tunnel submitted to him for examination every two hours, day and night. By night, if he was not in the tunnel, the samples were hauled up to his bedroom window in a basket. At last, in 1843—just one hundred years ago—the work was finished. A tunnel scarcely a quarter of a mile in length had taken eighteen years to build at a cost of eleven hundred and forty pounds for every yard of its length. It never served its intended purpose as a vehicular tunnel, for the sloping approaches were not constructed; but foot passengers used it for a time. Then, in 1866, it was incorporated in the East London Railway; now it forms part of London's Underground. Brunel's great work was, financially, a failure; as an example of sustained, dogged courage in the face of almost overwhelming odds it is a triumph. But, above all, it pioneered the approach to a new field of enterprise.

The civil engineer has to tunnel through all kinds of ground, from running sand to material which needs explosives to break it out. More often than not he has to deal with ground which is not self-supporting, which means that the tunnel must be provided with a permanent lining, or it would fall in.

In soft or loose ground it is the aim of the tunnel builder to get this permanent lining erected as early as possible. But the surrounding ground may not oblige him by waiting for him to build his permanent lining, however speedy he may be. So *temporary* support has to be provided to hold up the ground until the permanent lining can be built. Timbering was almost always used for this temporary support in the old days; (it is by no means out of use to-day). But the timbers have to be set up afresh for every short step forward and much time and labour are spent in doing so; a good deal of timber may be wasted, too. Now it occurred to Brunel, before he started the Thames Tunnel, that this repeated work of timbering could be avoided if the ground were given temporary support by a strong, movable framework which could be pushed forward bodily, every so often, as the driving of the tunnel progressed. There you have the principle of the tunnel shield. Brunel used his invention in the great work I have already told you about, but his practical interpretation of it was clumsy and not altogether successful. Its later development, however, has made tunnel construction in soft and loose ground much simpler and much cheaper than is possible with the old-fashioned method of timbering.

We have to thank James Greathead, more than any other man, for the tunnel shield as we know it to-day. Essentially, it consists of a steel cylinder, stiffened internally, and of a diameter slightly larger than the outside of the permanent lining of the tunnel. About half-way along the length of this steel cylinder powerful hydraulic jacks are fitted. These jacks are for pushing the whole shield forward a short distance—each "shove" is generally about 2 feet—as often as the miners have dug out enough ground at its front end to ease its progress. (You will understand from this that the shield is *not* an excavating machine—it is merely a travelling support for the ground outside the tunnel.) After every forward "shove," another short length of the permanent lining is erected *inside* the back end of the steel cylinder of the shield. Thus, the tail of the shield is continually sliding forward over the last length of permanent lining built under its protection, but it always overlaps the permanent lining by some amount.

If you've been able to follow this description of how a tunnel shield operates you may ask "What about the small space left between the outside of the permanent lining and the surrounding ground—the space of about two inches left behind by the skin of the shield as it slides forward? Is this space allowed to fill up naturally in course of time?" No, it is not; the result of such a thing might be settlement of the ground above the tunnel and damage to property on the surface. So the space is filled with liquid cement, forced in under pressure, each time the shield is moved. This cement soon sets as hard as stone.

Nearly all the tube railways in London, and scores of tunnels for all kinds of purposes in various parts of the world have been driven with the help of shields of the Greathead type. (Incidentally, mention of tube railways prompts me to point out that the longest railway tunnel in the world is *not* the Simplon with its 12½ miles, but London's Northern Line, on the Underground. From a point near Morden in the south to where it again emerges into the open near Golders Green in the north, this tube railway is in tunnel for a distance of just over 20 miles.)

Another powerful weapon in the tunnel engineer's armoury is the use of compressed air. Tunnels have frequently to be driven in water-bearing ground. By shutting off the tunnel workings from the outside air and raising the pressure of the air within them, the water in the ground can be kept out and the work of driving the tunnel carried through in the dry. The pressure required depends on the level, relative to the tunnel, of the source of the water which would otherwise cause flooding; but there is a limit to the pressure that can be used. With increase of pressure there is increased risk of compressed-air sickness; this is painful and sometimes dangerous, and tunnel miners know it as "the bends." It is caused by bubbles of nitrogen which, at the time of decompression (that is, when emerging from compressed-air) may be released from the blood just as bubbles of gas are released from a "fizzy" drink when it is uncorked—though on a much smaller scale. If these released bubbles of nitrogen do not get dispersed, but lodge in the blood vessels or tissues, trouble results. All concerned with compressed-air work are under a great debt to those doctors and others who, by experimenting on themselves under trying and even risky conditions, have discovered the causes of compressed-air sickness and shown what to do to prevent or relieve it. In practice a pressure of 40 to 45 pounds to the square inch above atmospheric pressure is about as high as you can go. (By the way, if ever you should enter compressed air don't expect to whistle. You can't.)

It seems ludicrous to think of railway tunnels as death-traps; but some of our forefathers did. Before the older Brunel had finished his difficult venture under the Thames, his son had completed the Box Tunnel (then the longest in Britain) on the main line between London and Bath. Its two miles were terrifying—too terrifying for many passengers to face—they preferred to break their train journey short of the tunnel, take a coach, drive the next two or three miles and rejoin the railway beyond. Perhaps their fears had been increased by an opponent of the railway company's plan to build the line. He had publicly expressed the opinion that "the monstrous, extraordinary, most dangerous and impracticable tunnel at Box would cause the wholesale destruction of human life." In spite of this gloomy forecast, the Box Tunnel is, like many other of the younger Brunel's works, still in service.

The most notable work of tunnel construction in this country in recent years is the Mersey Tunnel, which gives vehicular traffic direct connection between Liverpool and Birkenhead. Its main under-river section has an internal diameter of 44 feet and provides accommodation for four lines of traffic on a roadway 36 feet wide. At its lowest point the roadway is 148 feet below high water level in the river above. The length of the main tunnel—there is a branch entrance as well on each side of the river—is two miles. In the first twelve months of operation it was used by more than three million vehicles and the numbers were increasing each year up to the outbreak of war.

The traffic for which the Mersey Tunnel was designed to cater is, for the most part, motor-driven—a factor which increases the cost of modern underground roadways, for the exhaust gases from internal-combustion engines are, as you know, poisonous, and the provision of sufficient ventilation to prevent harmful concentrations of these gases is no small item in the bill. Of the total cost of constructing the Mersey Tunnel (about five and a half million pounds), ventilation accounted for approximately one million.

A road tunnel sometimes offers the solution of a traffic problem when no other satisfactory answer can be found. And that suggests the question: Are tunnels for road traffic desirable only when nothing else would really meet the case? Or, is it better to send traffic underground than to provide new surface roads or to widen existing ones, other things, including cost, being equal? (I am, of course, speaking of congested, built-up areas—nobody I imagine would want to burrow under open country!) But the same questions arise when it comes to choosing between a bridge and a tunnel for a wide river crossing. What would be your verdict then? These matters raise major issues of policy and planning as well as individual preferences.

Of tunnelling work carried out in connection with the war you will not expect me to speak in detail, although I could tell you a good deal. But tunnels have contributed much to our security by providing protection for all manner of people and all kinds of things in all sorts of places.

It is a far cry from the Thames Tunnel of 1843 to the Mersey Tunnel of 1934 and there has been a revolution in the technique of tunnelling during that period of nearly a century. Mechanical engineers, geologists, and those research workers who have studied the cause and prevention of compressed-air sickness have contributed to this. They have helped to apply tunnel construction increasingly to the creation of much that is essentially in our daily lives, with the result that it is now one of the most important means by which the civil engineer is privileged to serve his fellows.

SCIENCE IN NATIONAL LIFE

By E. C. BULLARD, F.R.S.

WHEN one listens to a talk about a particular application of science, to explosives, to textiles, to shipbuilding, or to some other practical subject, it is easy enough to see that the things talked about are interesting and important, but it is not so easy to see that they do not come about by chance because someone happens to have a bright idea. In this talk, the last of the series, I want to try to show that they come from the application of that great body of knowledge collected in the past 300 years and called "physical science."

Now what is physical science? It is the study of the behaviour of inanimate matter. The motion of the planets, the design of a steam engine, the production of petrol, the breaking of a wave on the seashore, the colours of dyes, and the interior of an atom, are all parts of physical science.

Such things may be studied for two reasons. We may study them because they are interesting, and we like to see the connections between apparently widely different things without any ulterior motive, or we may study them to achieve some practical end. The first is called "pure" science and the second "applied" science. Pure science is, I suppose, closely connected with the small boy's desire to take his father's watch to pieces. He does not, unless he is unusually optimistic, expect that the watch will go any better after he has taken it to pieces, he simple wants to see how the wheels go round, and to understand how it works.

The experiments made a hundred years ago by Ampère and Faraday to find out if there is any connection between electric currents and magnetism are an example of this kind of science without practical motives. Applied science on the other hand is based on an interest in solving some practical problem, such as how to design a better electric motor for a vacuum cleaner, or how to make sea-water drinkable.

Now it is obvious that you can't design a new motor without knowing a good deal about electric currents and magnets, and that the designer of the first electric motor. could not have started unless he knew a good deal about them. No one could have said 150 years ago that magnets had anything to do with providing power for sweeping carpets, or making telephones or indeed that they were of any practical use at all, except for picking up pins and making ship's compasses.

Ampère and Faraday showed that a wire carrying an electric current would move a magnet, and that moving a magnet near a wire would make a current flow in it. It is on these discoveries that the electrical industry is based.

The development of electric power and light at the end of the last century and of the immense industry that has grown up round it were absolutely dependent on these discoveries of Ampère and Faraday. Of course, if they had not made them someone else would have made them, but until they were made electric motors could not have been designed. The electric motor could never have been discovered accidentally, or by a sudden bright idea by someone who usually made steam engines. It could only be invented after someone had found the fundamental laws of electricity and magnetism. Until these laws were found it was not clear that they would be of any use, or in fact that they existed. I do not think that they could have been discovered from any other motive than pure curiosity about how nature works.

It is at first sight rather odd that a very practical matter like the electrical industry should be built on such unpromising foundations. Such a connection is, however, very common, in fact, almost universal. If you want to solve some practical problem you can usually only do it if you know where and how to start looking for the solution, and in any but the simplest problems you can only tell where to start looking if someone has done a lot of spade work first. There are many examples—the wireless valve couldn't be invented till someone had studied the behaviour of the electrons given off by white hot metals in a vacuum—but when, 40 years ago, J. J. Thomson and his students were making the fundamental experiments on these things, they can have had no knowledge of the direction in which their discoveries would be applied, though they may very well have felt that such advances in pure science would- bear practical fruit eventually.

I came across another rather topical example recently. I was talking to an astronomer who had become interested in explosives, and I said: "Well, anyway you are having a holiday from stars." "Oh, I don't know," he replied. "I find the work I've done on variable stars comes in very useful; after all, an explosive just after it has gone off is only a mass of hot gas, and so is a star—the theory is very much the same for both."

This solving of a problem by seeing its connections with more familiar things is at the root of much applied science; but such a connection is only useful if we know something about the more familiar thing with which our problem is connected. It is not helpful to say that an explosion is in some ways like a star, unless you know a good deal about stars.

I hope I have said enough to show the justification of pure science from a practical point of view. The first essential for a harvest of useful discoveries and inventions is a continual sowing of the seed of pure research. Pure research has largely stopped now, owing to the war. If it were not restarted there would be little immediate effect, but the practical discoveries of twenty years hence would have been cut off at their source.

The time required for a discovery in pure science to be

applied is longer than one might expect. It is ordinarily twenty to fifty years. The fundamental electrical discoveries which I mentioned just now were made between 1820 and 1835, but it was not until the eighties that the streets of London began to be lit by electricity. Again, Clark Maxwell—the first professor of physics at Cambridge—perfected the theory of electro-magnetism and predicted the existence of wireless waves in 1865 and Hertz first produced them in 1888, but practical wireless telegraphy did not come till 1900, or television and radio-location till the late thirties. The electron was discovered in 1896 but wireless valves were of no practical importance till 1915.

These time intervals are so long that there is, generally speaking, no possibility of making pure research pay for itself. The practical gains come from the fusing of the work of different men over many years into a coherent picture of how things work, most of the items being of no special value by themselves. The engineer who designs a ship uses a knowledge of the mechanics of water. This knowledge has been accumulated gradually during 250 years from the time of Newton to the present, or perhaps I should say during the 2,000 years since the time when Archimedes had *his* bright ideas while sitting in his bath in Syracuse. Clearly the men who accumulated this knowledge could not have made a living or have paid for their experiments directly by selling such knowledge from year to year as it was acquired. We therefore see that pure research, although essential for practical ends is not self-supporting, and how it is that pure science, although one of the best investments that a country can make, is at the same time perpetually short of money. The trouble is that it is, as business investments go, a little slow in maturing, and also that the financial benefits cannot usually come directly to those practising or supporting it. Copernicus could not patent the discovery of the motion of the earth round the sun, nor could Newton sell the law of gravitation, yet, in the long run, their work was immensely valuable. It enabled more accurate predictions to be made of the positions of the sun, moon and stars, and thus made the navigation of ships more certain. Once the motions of the heavenly bodies were understood it became a matter of applied science to make the best use of this knowledge, and this applied knowledge was saleable. During the eighteenth century a much increased demand for instruments for navigation grew up, based on the new methods and depending on the astronomical discoveries of the two preceding centuries. The government offered a £20,000 prize for a clock that would keep time on board a ship and firms of instrument-makers set up in business to supply this and other wants of sailors. As some of these firms still exist we may suppose that they have found the making of sextants and chronometers profitable.

From this one would expect that pure science would only flourish in a fairly prosperous country. In a country, that is, that can afford to lay aside a certain proportion of its resources each year, and use it for work that will not bear fruit for a number of years, and much of which will never be of any use at all. Until recently this laying aside was done largely through private benefactions to universities. However, as the simpler parts of science get worked out and attention is directed to more and more deeply-lying parts, the complexity of the equipment required gets greater and the cost increases. The giant telescope that is to be erected on Mount Palomar in California will cost over £1,000,000, and the atom-splitting cyclotron at the California Institute of Technology cost several hundred thousand pounds. These are sums that are, except in occasional instances, quite beyond the reach of private benefactions, and progress in science has come to depend more and more on government grants derived from taxation. It seems likely that if this country is to keep up with the startling developments in the United States and in Russia this process will have to go farther still. This raises difficult questions of how to retain the traditional and necessary freedom and independence in the development of science, and at the same time secure efficient administration in the spending of large sums of public money. I will leave it to you to discuss this in detail. I must return to my main theme, the relations between pure and applied science.

I have spoken of the contribution of pure science to practical affairs and to industrial development, now I want to talk about the reverse connection. Most advances in pure science depend on new methods of doing things and these new methods are provided by advances in applied science. There is thus a kind of alternation between pure and applied science, each depending on the other for its advances. For example, a hundred years ago the simplest electrical experiments were very difficult to carry out, because of the absence of elementary requirements like electric wire. Faraday, in his diary, describes how he wound his magnets with milliners' iron wire which was, and for all I know still is, used for stiffening ladies' hats. The insulation was provided by winding string between the turns of wire. As I have explained, these experiments of Faraday's led in the course of 50 years to the electrical industry. With the development of that industry things like insulated wire, magnets, dynamos, motors and accumulators became commonplace, to be bought in a shop. This immeasurably simplified all experiments involving electricity, and rendered possible the fundamental researches of J. J. Thomson, and others, which led to the discovery of the electron. These experiments in turn led to the invention of the wireless valve, which has itself laid the foundations of the great and expanding radio industry— and to my talk to-night. This industry has made valves a commonplace in the same way that earlier electric developments provided lamps and motors for all. This plentiful supply of valves has made possible the experiments of the last fifteen years on the structure of the atom and has led to an enormous increase in our knowledge of how atoms work.

It is quite possible that this knowledge will find some quite unexpected application and be the key to the growth of some industry, which will in its turn make easier a new attack on yet more fundamental properties of matter.

I now come back to where I started. The practical advances about which the other speakers in this series have talked do not come by chance because someone happens to have a bright idea. The bright ideas must be there, but no one can have them without a background of knowledge collected by others. No one can start having ideas about, shall we say, how to make better dyes until they know what sorts of arrangements of atoms give coloured compounds and what happens if you add a nitrogen atom here or a carbon atom there to the dye molecule. I happen to have chosen the chemistry of dyes as an example; but this point was well brought out in earlier talks on Plastics and Explosives.

There is just one other thing: the development of science requires not only the right men, but also the right circumstances, and given the right circumstances the men will usually appear. Both in pure and in applied science you need not only the individual genius, the Newton, the Faraday, or the Rutherford, but also the right surrounding circumstances. For such advances a country must be rich enough to make an investment that will not show a profit for many years. For such advances a country must be large enough to provide an adequate supply of rank and file scientists as well as a few great men. It must be sufficiently industrialized to provide the means, and to some extent the incentive for advances. Without all these neither pure nor applied science will flourish. We can see this if we consider why this country, the United States, and Germany have during the last fifty years contributed such a large proportion of the world's scientific progress, and countries such as Spain and China have so far contributed relatively little.

A scientific discovery is not a thing in itself which develops separately from the rest of the life of the country, it depends very intimately on the wants and habits of ordinary people, on the kind of government they have, and on the educational and financial system they support or tolerate. We are no more intelligent than our ancestors, but we have learnt more about nature in fifty years than was learnt in the first 1,500 years of our era. To state precisely what are the conditions for a healthy development of science is a difficult thing: the conditions that give that combination of curiosity, self-confidence and patience that make a scientist are subtle and elusive, but it is at least certain that practical discoveries do not happen of themselves, they depend on threads running right through the fabric of the country's life.

THE END